ROBOTICS AND INTELLIGENT MACHINES IN AGRICULTURE

Proceedings of the First International Conference on Robotics and Intelligent Machines in Agriculture

October 2-4, 1983
Curtis Hixon Convention Center
Tampa, Florida

Published by:
American Society of Agricultural Engineers
2950 Niles Road
St. Joseph, Michigan 49085-9659

ii

TABLE OF CONTENTS

PREFACE

Man has been the primary source of intelligence in his agricultural systems since he began farming. But microprocessor technology has progressed and costs have fallen to the point that the widespread application of intelligent machines in agriculture is imminent, and the application of robotics now seems possible.

This proceedings results from the first International Conference on Robotics and Intelligent Machines in Agriculture, organized by members of the Agricultural Engineering Department of the University of Florida, and held as one track of Robotech '83 in Tampa, Florida, October 2-4.

The objectives of this conference were to review current research related to intelligent machines and robotics in agriculture, to project future applications of robotics in agriculture, and to stimulate new research in these areas. The papers by Anson, Gerrish and Surbrook, Key, Kawamura, Ito, McClure, Grand d'Esnon, Johnson et al., Mancaster, Sullivan, and Krutz and Mailander give a fascinating insight into recent and current research. Papers by Tesar, Anson, Krutz, Tutle, Siegel, and Orrock and Fisher present diverse glimpses of the future. Of particular interest to Florida was the emphasis on robotic citrus harvesting in the papers by Tutle, Orrock and Fisher, and Coppock. If attendance and interest exhibited by the audience is any indication, it seems assured that research and new applications of robotics and intelligent machines to agriculture have been stimulated by this conference.

Richard C. Fluck
Proceedings Chairman
October 28, 1983

INTERNATIONAL CONFERENCE ON
ROBOTICS AND INTELLIGENT MACHINES IN AGRICULTURE

October 2-4, 1983
Curtis Hixon Convention Center
Tampa, Florida

Co-Sponsors:

American Society of Agricultural Engineers (T-3 Committee for Instrumentation and Controls)

Agricultural Engineering Department, Institute of Food and Agricultural
Sciences, University of Florida

City of Tampa, Development Department, Office of the Mayor

Program Committee:

L.N. Shaw, Chairman
Agricultural Engineering Department
University of Florida

G.W. Isaacs
Agricultural Engineering Department
University of Florida

W.D. Shoup, Publicity
Agricultural Engineering Department
University of Florida

M.A. Purschwitz
American Society of Agricultural
   Engineers

R.C. Fluck, Proceedings
Agricultural Engineering Department
University of Florida

J.D. Whitney
AREC, Lake Alfred
University of Florida

G.W. Krutz
Agricultural Engineering Department
Purdue University

G.E. Coppock
Florida Department of Citrus
AREC, Lake Alfred
University of Florida

D. Tesar
Center for Intelligent Machines
   and Robotics
University of Florida

D.K. Kanoy
Conference Planning
University of Florida

Technical Advisors:

ASAE T-3 Committee for Instrumentation and Controls

W.F. McClure   J.R. Lambert

B.W. Mitchell   J.L. Steele

v

ROBOTICS: ECONOMIC, TECHNICAL, AND POLICY ISSUES[*]

Delbert Tesar

Industrial robots are now a reality, and they are having an impact in the field of automation. But what is their potential economic benefit? What are some of their future applications? What are some of the limitations of the existing technology, and is our present R&D effort sufficient and on target? These questions are addressed in this brief listing of issues facing those involved in the development of robots.

Over the past several years, I have analyzed the conditions affecting manufacturing in the United States, documented our relative competitiveness—not only internationally, but especially in our home markets—and pointed to some basic factors associated with our weakening trade capacity.

Manufactured goods represent 60% of total U.S. trade activity, and mechanicals (automobiles, office machines, cameras, etc.) represent 75% of that total. Yet only 6% of the nation's research and development money is spent to support this important portion of U.S. trade. Furthermore, only 0.7% of our federal R&D in manufacturing goes to enhance technology associated with mechanical manufacturing. That has led to a weakening of our trade capacity in manufactures (i.e. value added goods).

The imbalance between the amount of mechanicals sold by this country and the amount of R&D money invested in the manufacturing process is at the core of our trade weakness in that field. This is most vividly demonstrated by the fact that the 20 worst deficit generators in the mechanical trades category created a $34 billion trade loss in 1978, equivalent to this country's loss in oil for that year. The figure is likely to approach $50 billion for 1982.

In 1978, Japan had a $63 billion surplus in manufactures and Germany had a $49 billion surplus, primarily in civil sector goods. These countries have strong national priorities to vigorously pursue vital technologies such as robotics for manufacturing. By contrast, Australia had a $10 billion trade deficit in manufactured goods (an equivalent, comparatively, of $200 billion for the U.S.), with negative trade ratios approaching ten to one in several categories. It is important to realize that 80 percent of the manufacturing in Australia is mechanical in nature. In most industrial countries, it is estimated that merchanicals represent 70 to 80 percent of manufactured goods. It is in the field of quality (precision) mechanical manufacturing capable of rapidly following a market (the customized product) for a competitive price (more for less) that robots can play a major role. Robotics—and the concept of the intelligent machine—is an emerging technology which

---

The author is: Director, Center for Intelligent Machines and Robotics, and Professor, Mechanical Engineering Department, University of Florida, Gainesville, Florida.

[*]Excerpted from: "The Robotics Race is On," D. Tesar, Robotics World, January, 1983, pp. 32-33. "Robotics: An Economic Issue," D. Tesar, Windows, Texas A&M Experiment Station Journal, October, 1982.

can be used as a catalyst to strengthen civil sector manufacturing. There-fore, the field is of the highest importance to a nation's economic well-being.

Historically, American products were considered to be of good quality. During the 1950's, U.S. machine tools were considered the best in the world. Now, not only is Japanese equipment competitive in quality, but also in price--evidenced by the fact that they now produce 45% of the world's robots. Robotics--the concept of the intelligent machine--is an emerging technology which can be used as a catalyst to offset our present lack of competitiveness in civil sector manufacturing. Enhanced manufacturing pro-ductivity and reduced hazards to operational personnel performing dangerous tasks are two primary pressures which govern present and future applications for robotics.

It's important to realize that 25,000,000 people are now employed in manu-facturing in U.S. industries, of which more than 10,000,000 are performing manual functions on a semi-repetitive basis. Today, it is estimated that 5,000 robots are operating in the U.S. This represents a penetration of not more than 5 in 10,000. Other countries are outstepping the United States. For instance, increasingly Japanese plant development is moving towards the manless factory or the factory without lights. Similar advanced factor development is taking place in northern Europe.

At the same time, the United States' annual robot research and development budget, some $16.5 million, is smaller and much less cohesive than that of Japan and France. Advanced factory development similar to that of Japan is taken place in northern Europe. The technology of intelligent, flexible machines as represented by robots is perhaps the central technology involved in these efforts. It is apparent that expanding our activity in the field of Flexible Manufacturing Systems (FMS) will require a major cohesive effort by all parties in the entire U.S. technological community.

During the past two years, the Russians have established robotics develop-ment and applications as a top national priority. Several scientific acade-mies and institutes have established research teams comprised of 20 to 100 members each. The Paton Electric Welding Institute in Kiev is the larg-est. The Soviet Union's current five-year plan calls for the construction of 40,000 robots. Progress is occurring in Russia, since robotics manufac-turing increased 88 percent in 1982 to a total of 3,600 units. The range of application is very similar to those in the West. The Soviets have been major buyers of Western robots, including 50 Unimation robots and several French welding robots by Sciaky. The Soviets are known to have theoretical strengths in artificial intelligence (Leningrad), vision (Kiev), and struc-tural kinematics (Moscow). Their primary weakness derives from their lack of proven component technologies, including compact microprocessors, en-coders, servo-valves, precision hydraulic and electrical prime movers, and other robotic devices. Overall, the Russian technology is believed to be a decade behind that of the West.

Among the Communist bloc countries, Bulgaria represents a major center for robotics development outside of Russia. A factory for robotics is being established in East Germany and cooperative ventures among the Eastern bloc countries in robot welding cells are being pursued. An extensive effort to standardize robotic systems, software, and components is now being under-taken by principal decision makers in the COMECON countries. This effort includes a world-wide survey of the state-of-the-art which will result in the listing of technical requirements to meet projected needs over the next decade. On this foundation, component and system technologies will be dele-gated to groups in the various countries for research and development and manufacture. The goal is to create a rapid, but orderly, expansion of robotics technologies in the Eastern bloc. The importance of this effort

2

was highlighted by a trip by the late President Brezhnev to East Germany in 1981, to organize that country's portion of this cooperative venture.

The French government recently has added robotics as a new element to its national strategy for enhanced technological vitality. Its minister for research and industry has proposed to spend $360 million over the next three years for robotics. This program includes $36 million for equipment at research laboratories; $325 million for research grants to develop advanced robots; and four to five geographically distributed robotics research centers. The minister states that the goal is to boost French productivity by seven percent annually during the next decade. This is twice the increase expected in the United States.

Beginning in 1945, the technology of robotics was derived from the urgent needs associated with handling nuclear materials. Since 1970, however, use of the robotic manipulator capable of duplicating complex human motor functions in industrial applications has made significant in-roads in the field of automation.

As we look ahead to the next generation of robot systems, we see not the type of human imitations made famous by Hollywood, but machines capable of precise measurements and movements. The human system performs its functions because of unsurpassed hand-eye coordination. Nonetheless, the human system is probably a very poor model on which to base the design of a machine which will perform precise mechanical operations under load. This fact has not been widely recognized and, to a great extent, slows the scientific community's response to this valid need. No human hand is capable of precision measurement or is capable of precision machining operations under load. Therefore, the human model for robotic manipulators is adequate for the simple repetitive tasks now being pursued in industry such as pick-and-place, spot welding, spray painting, surveillance, and unloaded assembly. A precision system capable of high loads in real-time operation could perform precision machining without jigs, precision inspection during operation, multi-level operations under CAD/CAM control, and force fit assembly.

Most manipulator arms are accurate to about 0.0505 inch, which is insufficient for many precision operations such as assembly operations and spatial motion functions such as laser cutting and welding operations. Precision requirements will intensify as robotic systems become smaller and are applied to such functions as microsurgery. Generally, deformations under small applied loads in existing commercial systems substantially exceed the static (or repetitive) precision level they can achieve. In addition, the operating environment—temperature, loads, shocks and vibrations—cannot always be considered invariant. Hence, future robotic systems must be much more adaptable to "control-in-the-small," and real-time modeling to meet this second higher level of precision must be developed in the next generation of robots.

Examination of robotic technology shows certain important facts. The industrial robot arm may cost up to $100,000. In contrast, the specially designed space shuttle manipulator development and deployment cost was $100 million. It's evident that unusual technology in this field is extraordinarily expensive.

Ocean floor activity illustrates the usefulness of robots to perform hazardous tasks. With 19% of the world's crude oil coming from offshore sources, the ocean floor has great economic potential if a means of tapping that potential can be found. Costs for offshore drilling efforts are high. The Mobil Oil Condeep drilling platform costs $1.3 billion to handle 42 wells, some as deep as 9,300 feet, and operates in 500 feet of water. Future systems planned by Exxon will be submerged and operated at a depth of 5,000 feet. Remotely Controlled Vehicles (RCV) become economical at these depths. For example, at 1,000 feet, a diver may work for 10 minutes at a total cost of $100,000, while a remotely controlled craft costs $3,500 per day. Today's RCV's are relatively simple, but they can inspect structures, pipelines and cables; place and recover instruments; set explosives; clear fouled lines; and survey under ice.

Effective development of robotics technology demands integration of disciplines that are frequently considered separate. The primary marriage is of mechanical and electrical engineering, although many other technologies are involved as well. There appears to be insufficient effort towards combining man with machine to enable the uniform penetration of this emerging technology into the workplace. The best opportunity in the immediate future is to balance human and machine capabilities for many applications. As the machine technology improves, less will be asked of the human and more of the machine. This man-machine approach allows the most rapid penetration of the manufacturing market with the near-term technology, allows a gradual and natural transference to more machine-oriented systems, and allows a minimum disruption of the manufacturing workforce.

The time of the intelligent robot is upon us. Virtually every industrialized society has significant research and development in robotics activity as well as a first generation of applications. As the number of applications increase, the demands on the technology will result in carefullly integrated systems increasingly made up of modules and scaled more effectively to the task. Miniature robots will be essential for micro-surgery and the assembly and repair of micro-circuits. No longer will "add-on" technologies suffice--the present disciplines of mechanical and electrical engineering will tend to merge into what the Japanese call "mechatronics." Society, in general, will accept the image of manufacturing as a high technology field and reap the benefits of less-expensive, yet higher-quality goods.

I conclude that: robots are an important part of the future for manufacturing, and that future robots will perform functions beyond today's concepts of the machine; robots will augment human capabilities (not only displace the operator and thus de-skill our workforce); robots cannot be oversold to the young student or researcher, but present robot technology can be oversold to decision-makers in industry; and that our present R&D level in the field is not only insufficent, off-target and incohesive, but that other nations are accelerating their development to our future disadvantage.

4

TECHNOLOGICAL TRENDS IN AGRICULTURAL ELECTRONICS

James H. Anson
Member ASAE

Ladies and Gentlemen, it is an honor for me to be able to visit with you as part of ROBOTECH 83. I am looking forward to attending the entire program and, hopefully renewing acquaintances with many of you that I do not get to see very often.

When Gerry Isaacs asked me to speak, I told him that I knew very little about robots — he said that was OK, because there would be plenty of people at this conference who do. We also discussed briefly how electronics were penetrating virtually every aspect of food production and distribution — planters, tractors, combines, forage harvesters, sprayers, spreaders, dairy operations, meat operations, grain elevators, processors, on-farm management systems, laboratories, irrigation, cotton harvesters, and the list goes on and on. For this presentation we agreed that I should try to give my views of the technological trends in the farm equipment industry.

I can assure you that any forward-looking predictions I make are entirely my own, based on the 15 years I have spent in the agricultural electronics field, and may not even represent the feelings of some of my associates. In addition to that disclaimer, even though I consider myself an engineer, there must be some small chance that I could be wrong in my predictions. To cover that possibility, I have asked the publishers of these proceedings to list me as an economist.

We need to take a brief look at where we are today in agricultural equipment, and when we do we find that electronics are being routinely used to improve productivity and efficiency, improve safety, reduce operator fatigue, and to provide much needed operator information. In addition, closed-loop controls are becoming more prevalent. Since beginning with capacitance-type moisture testers and planter monitors in the late 60's, many agriculture machines have been equipped with electronic instrumentation. Let's take a look at a few of these instruments:

1. Electronic planter monitors for verifying that seeds are accurately and consistently placed in the ground.

 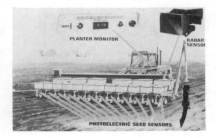

Fig. 1 — Original Planter Monitor   Fig. 2 — Dj Planter Monitor and Sensors

Fig. 3 — John Deere Planter Monitor

2. Combine Monitors — — Since a combine is a "Factory On Wheels", these devices assure of properly sequenced operation of shafts and other mechanisms.

Fig. 4 — Combine Shaft Monitor

3. Grain Loss Monitors which statistically evaluate the amount of unthreshed grain and inform the operator.

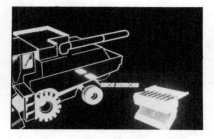

Fig. 5 — Dj Grain Loss Monitor   Fig. 6 — Grain Loss Sensor

4. Tractor Instrumentation – – To monitor and display vital tractor information to the operator.

Fig. 7 – IH Tractor Instrumentation

Fig. 8 – AC Tractor Instrumentation

Fig. 9 – J. I. Case Tractor Instrumentation

Fig. 10 – J. I. Case Tractor Instrumentation
(On Tractor)

5. Electronic performance monitors which determine machine performance on wheel slip and acres per hour. These instruments routinely use radar for true speed and distance measurement.

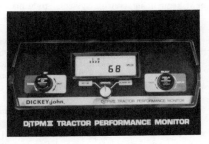

Fig. 11 – Tractor Performance Monitor

6.   Planter controls which permit operators to change population on-the-go.

Fig. 12 — John Deere Population Control

7.   Closed-loop control to ensure that the proper amount of liquid or granular material is being applied.

Fig. 13 — Dj Sprayer Control

8.   In addition to the instruments shown, there are also in use automatic load controls which vary vehicle speed depending on load conditions, hitch control systems for uniform loading, implement sequencing control systems, transmission controls, and new types of servo valves.

With all of these instruments we find a rising degree of customer confidence in the ability of electronics to enhance the operation of their equipment. We also have a "next generation" of society coming that will be computer oriented and hungry for information — insistent on efficiency, comfort and relief from tedious, repetitive tasks. This new generation will think a slide rule was some form of ancient musical instrument.

Well, what comes next? As I mentioned earlier, predictions are difficult and risky, particularly so if you try to identify a time frame when events might occur. What does "occur" mean — well, to me it means when commercial success is achieved. Sometimes we tend to lose sight of the fact that technology and commercial success are interrelated. Let's define technology as — the room full of tools currently available to be applied to solve today's needs and the research work going on so that a bigger room full of tools will be available for tomorrow's needs. Let's define commercial success as the sale of useful products to fill real customer needs resulting in profits to all involved in the stream of utilizing technology to convert nature's raw materials into products that fill those needs. If you accept these definitions, we can conclude that technology and commercial success are forever dependent on each other. Commercial success funds technology and technology provides the tools by which commercial success can be achieved.

So, why all this dissertation on technology and commercial success — so we are on the same wavelength trying to predict a time frame when commercial success will occur, because technology precedes commercial success.

Let me give you an example, and I am reading from an article which I will identify after I read you this portion, "Will the farmer of the future be able to sit on his front porch while directing all his farm work? Will it be possible to sit in an office in Chicago or New York and direct the operation of fleets of tractors through the world? Will it be possible by these methods to operate farm properties in both hemispheres and gather harvests in practically every month of the year? What are the possibilities of radio control in housework, industrial work, transportation, and especially in warfare? These are a few of the unanswerable questions with which the weird spectacle of a driverless, yet perfectly controlled, tractor excites the imagination. A new field of possibilities has been opened by the new International Harvester Radio-Controlled Tractor of 1934." This article from the 1934 Century of Progress Show in Chicago. I also found an article on a 1914 tractor equipped with optional self-steering. Commercial success has not yet been achieved with either, even though enough technology existed then to make those machines.

For a good look into the potential future for agricultural electronics on farm equipment, one only needs to read the many trade publications available, including our own ASAE magazine, for there are many good predictions by recognized authorities on agricultural equipment. Let me give a few examples:

September, 1981 issue of AGRICULTURAL ENGINEERING — "Automatic Guidance For Agricultural Machines" by Lowrey Smith and Robert Schafer.

Highlights of this article are — Automatic guidance is feasible and should provide a desirable option on tomorrow's field machines. Also — ideally, automatic guidance systems should be completely versatile — the guidance function should not be limited to particular field operations or row patterns.

Such a system requires a spatial position — sensing system independent both of traditional machine operations and of field installed guidance devices such as furrows, standing crops, buried wires, etc.

Another example — September, 1983 MACHINE DESIGN — "Agricultural Equipment — Cash Squeeze Slows Development" by Richard T. Dann, Staff Editor.

Highlights are — A depressed market has slowed the technical development of ag equipment but advances are being made in efficiency and productivity and are on the horizon in automatic guidance and control. (Here you can see the relationship between technology and commercial success.)

I guess the word "horizon" must mean nearer than future but farther away than tomorrow. The article also says the main feedback control used for tractor/implement draft control is a concept that was commercially introduced in 1939. It says evolving agricultural practices will require greater machine control and farmers will need better performance, sophisticated information, and more automated controls. This includes minimum tillage, controlled traffic and equidistance plant spacing. Also — automatic steering aids would be a major stress reliever in cultivating and planting. This same article talks about improved hydraulics and electronics and mentions a production tractor that has an electronic display and control center that measures hitch position, operates an electronically controlled power-shift transmission and monitors tractor functions utilizing 16 warning indicators — it also sounds a warning horn in the radio speaker when malfunctions occur.

In the August, 1983 issue of AGRICULTURAL ENGINEERING "Soil Tillage, The Challenge Of Diversity" by Mariana Pratt — shows an 1860 concept of a wide frame carrier similar to a research wide frame carrier in use today. The article stresses the importance of controlled traffic. It mentions that in field equipment added efficiencies are being achieved with electronics and computer-sensor interfaces and this utilization is still in its infancy. It says various researchers seek to provide practical field guidance systems to keep equipment on established roadbeds. This article also mentions a 1963 challenge from USDA — "You must supply the best possible equipment for the production of a certain crop in a specific field of a farm in a given region" — a difficult task conflicting with volume production and universal distribution, but one which will become increasingly justified (once again — commercial success enters the picture). And, this article also says — new concepts which utilize new field equipment design and the latest in electronics may help that dream soon to become a reality.

So far we have heard the terms tomorrow, future, on the horizon, and soon.

Another article in the same August, '83 issue — "What's BEYOND The Horizon In Field Equipment" by Clarence Johnson and Bob Schafer. Highlights include the definition of custom prescribed tillage, CPT, — I like that term — a concept of applying a prescription of tillage practices based on knowledge and specifications of specific crop needs, soil attributes, climatic influence and any other factors that may influence management decisions related to soil tillage. Thus the concept is not limited to specific crops, soils, climatic regions and machinery systems — and the article continues, as we look beyond the horizon automatic control systems will become integral components in future field machinery systems. Finally, in the same article, an era of automatic controls in future systems may produce results that dwarf those produced by the era when mechanical power replaced manual power.

And, one last example in DIESEL PROGRESS, September, 1983 issue, a market decision article written by Rob Wilson around interviews with a vice president of engineering of a major farm equipment company. Highlights of this article regarding engine development includes use of electronics to accomplish energy management functions (I like that phrase), monitoring and controlling the process and bringing much higher efficiency throughout the drive train to ensure that the output is optimized for various conditions. It also points out that all machines, agricultural, construction and industrial will be far more versatile and contain more on-board electronics and sophisticated hydraulics. It says that the management of energy will be very important, the operator will be better informed, he will be part of the decision process but will probably not make all of the decisions. Due to better sensors and better monitoring, machines will last longer and do a better job for the customer (fill a need).

As you can see from just a few examples I have cited, that a great deal has been said by recognized authorities — and it all seems to add up to encouraging times ahead for electronics in agriculture. The technological trend is more electronics, for information and control, with more sophisticated and reliable hydraulics — continuing to supply the muscle. Now, let me try to apply time frames for commercial success to some of the technology I have mentioned.

Fig. 14 — Future Directions To Ag Electronics

Between now and 1990, I only see wider spread use of the present day systems. There will be more bells and whistles, improved sensors and more closed-loop controls, with some form of electronics appearing on every major piece of farm equipment. Since tractors achieve their value by pulling a variety of implements, chances are good we might see a more universal method of tractor-implement communication.

Fig. 15 — Generic Tractor

In block diagram form, we might see something like this —

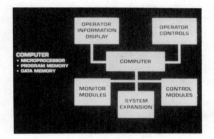

Fig. 16 — Block Diagram

One of the bells and whistles might well be voice synthesis or recognition.

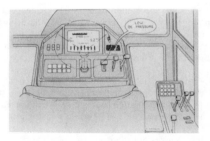

Fig. 17 — Word Synthesis

Tractors in this time frame are likely to include some or all of the following:

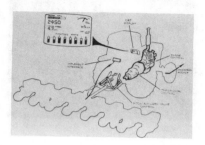

Fig. 18 — Tractor w/Engine Control, Transmission Control, Hitch Position Controls, Radar, More Universal Type Displays and Electronic Implement Interface Capabilities

And this next figure illustrates the variety of implements that could have their sensors talk to the tractor through the addition of modules.

Fig. 19 — Tractor With Implements

What about after 1990? Well, there are numerous technical challenges remaining which, hopefully, will keep most of us employed. We need more work on identifying the relationship between physical parameters and measurable electrical parameters — fertility, soil strength, yield, depth, height, are a few examples. Once these relations are identified, suitable transducers (sensors) must be developed that are accurate, reliable and of reasonable cost.

In addition, strategies for altering machine operation, whether it be combines, tractors, forage harvesters — whatever — must be properly determined and machine performance, soil and crop conditions must be converted into information that is relevant and meaningful to the operator. This information must complement the knowledge and intuition of the farmer.

No matter how you slice it — sensors are the name of the game. It's not good practice to build a penthouse on top of a crumbling foundation, and sophisticated monitoring and controls built around poor sensors generating inaccurate information is also a poor practice.

I see enough of these technical problems being solved to take a large step forward, between the years 1990 to 2000 toward what I consider to be the farm of the future, a farm which will utilize custom prescribed tillage, custom prescribed planting, custom prescribed spraying and spreading, custom prescribed harvesting and centralized farm management. I feel that commercial success in these areas can be achieved in this time frame with or without practical automatic guidance-field position indicator, which still remains a missing link. I think that some practical solutions to indicating field position will be prototyped within the next 10 years with commercial success possible between the years 2000 and 2010. I hope I am here to see it — for then everything falls into place for the farm of the future — everything that is done in the field prior to harvest will be measured and recorded. The actual yield will be measured and recorded. The results can be analyzed on the central farm management system, integrated with other variables such as rainfall, temperature, prices of seed, fertilizer, fuel and chemicals, and for the next year, the equipment could be pre-programmed based on the previous year's results. This measurement, recording and pre-programming could be transmitted to the farm management system, but most likely will be recorded on tape, analyzed during off season, pre-programmed into a new tape which can be inserted into the machine control system causing it to react efficiently at various positions in the field. As the years pass, the data base grows and efficiency becomes routine. And so, from something electronic like this in the late 60's.

Fig. 20 — Old Tractor w/Monitor

We progress to something like this in 50 years —

Fig. 21 — Tractor w/Radio Waves

I think that the operator will remain in the cab, available if needed to make critical decisions during machine operation, but the prime mover would qualify as a robot —

Fig. 22 — Closed-Loop Farm Management System

An intelligent machine, re-programmable, relying on a host of sensors for information and ranging accurately through a variety of fields adjusting itself as necessary for the highest level of efficiency and carrying its operator in comfort, supplying him with management information, relieving the operator of repetitive, mundane tasks, and retaining our reputation as the world's number one producer of food.

Thank you.

REFERENCES

1. Brochure. 1934. Century of Progress Show — Chicago. 34—35.

2. Dann, Richard T. 1983. Agricultural Equipment — Cash Squeeze Slows Development. MACHINE DESIGN. September. 42—51.

3. Johnson, Clarence and Robert Schafer. 1983. What's Beyond The Horizon in Field Equipment. AGRICULTURAL ENGINEERING. August. 10—12.

4. Pratt, Mariana. 1983. Soil Tillage, the Challenge of Diversity. AGRICULTURAL ENGINEERING. August. 6—9.

5. Smith, Lowrey and Robert Schafer. 1981. Automatic Guidance For Agricultural Machines. AGRICULTURAL ENGINEERING. September. 12—14.

6. Wilson, Rob. 1983. New Diesel Engine Development Cycle Soon To Begin. DIESEL PROGRESS. September. 38—40.

FUTURE USE OF ROBOTS IN AGRICULTURE

Gary W. Krutz
Member ASAE

Three factors are significantly changing our farms and businesses. They are world competition, consumer attitudes and technological advances, especially in computer capabilities. Stiff foreign competition (evidenced by steel and automotive industries since 1979) has now become prevalant in agribusiness within the last two years. The U.S. tractor industry is seeing a decline in its market and market share while the U.S. farmers' profits continue to erode to 10 years lows.

Also, the world is in an era of applying technology at faster rates than ever before. Applying high tech scenarios have been discussed in recent Wall Street Journal articles (Austin and Beazley, 1983; Ingrassia and Darlin, 1983; Hynowitz, 1983). Donald Grierson, G.E., elaborated on world-wide competition and the need to automate, "The advantages that your world competitors will have if they automate and you don't will be so significant that it will become a question of survival for many companies." This trend towards automation is now evident in the agricultural machinery marketplace with the majority of tractors sold in the U.S. being imported. Agricultural crops that have high labor usage, such as grapes, apples and oranges, are now being imported from Chile, New Zealand and Brazil, respectively. The labor rates in those countries are significantly lower. Austin (1983) discusses the labor loss situation citing comments from labor that a few jobs in an automated firm are better than none at all.

Automakers are using robots to increase productivity and reduce costs with estimated usage of more than 5000 by 1985. Farmers and agribusinesses are considering similar adaptations of technology.

Part of the lack of movement towards automation is being blamed on education. Japanese students have been shown to have the highest math and science scores in the world, play two musical instruments, do twice as much homework and have a 90% high school graduation rate compared to 75% in the U.S. This has provided highly motivated youth for their science and high technology fields. However, these pressures do have negative aspects. A Florida school Superintendent Raymond Shelton (Brechenridge, 1983) points out that the pressure in Japan is so high that it creates a high suicide rate among high school students. Also, once admitted to college, the Japanese students slow their efforts and Americans surpass them upon graduation.

All of this competition is good because it keeps our people, farms, and industry from becoming complacent and continually aware of future trends. But, recently we slowed our annual increases in industrial research and development spending. During 1980 and 1981 real R&D spending increased 6.5% per year; in 1982 and 1983 the increases were between 2 and 3%. Historically in the 1950's and 1960's the rate of increase per year was 8.5%.

The author is: Gary W. Krutz, Associate Professor, Agricultural Engineering Dept., Purdue University, W. Lafayette, Indiana.

McGraw-Hill (1983) discusses the effect of this declining rate. "This slippage, however, apparently produced at least one positive side effect: It galvanized many U.S. businesses into reinvigorating R&D programs.

"For example, from 1967 to 1975 foreign auto makers increased their share of the U.S. market from 9% to 18%. U.S. car makers, meanwhile, were hiking their R&D budgets 7.5% a year on average. Since then, however, U.S. auto companies have boosted R&D budgets an average of 11.5% a year. And even though U.S. motor vehicle sales declined some 4% last year, producers still raised R&D budgets a bit. More dramatically, despite the fact that 1982 sales were off 6% from 1977, auto industry R&D spending between 1977 and 1982 zoomed 48%.

"Thus, judging from the relatively generous allocation of corporate funds to R&D in recent years, it seems that U.S. firms are extending their planning horizons. In other words, the era in which American business was preoccupied with short-term results may be giving way to a period in which investment for the long haul comes back in vogue."

History has shown us a lesson to look at. During the late 1970's U.S. industry, in real dollars, reduced their R&D spending and lost their markets to worldwide competition.

TRENDS AFFECTING AGRICULTURE

In a recent article the changes in farm structure are discussed by Krutz (1983). Consumers are the key to watch, as they will affect the direction agricultural technology takes. They have and will continue to determine the methods of food preparation. One significant population trend is the growth of the older and aged segment. The Social Security Administration has projected a decline to two workers for every retiree by the turn of this century. Along with this trend will be the increased income level of all workers. Both of these changes could lead to increased demand for quality in all items we purchase.

How does this affect food and agricultural machinery? For example, the demand for higher quality food will parallel a demand for more nutritious food. Items such as vine ripe tomatoes (red and tasty) may replace those picked green and hardy for shipping and chemically ripened. From this crop one can surmise an increase in locally and environmentally grown vegetables.

U.S. consumers will demand quality foods in the future especially when they reduce their caloric intake while dieting. Dieters are interested in vitamins, nutritional quality and the chemical processing of the food they eat. Other trends Americans see that will affect agriculture include shorter workweeks and increased educational levels.

Worldwide competition for all markets including agricultural products is becoming more competitive. This is one reason why U.S. farm income has dropped as seen in Fig. 1.

The major reason for the decline in farm income is the changes in foreign currency exchange rates. During the period 1979 to 1983 we've experienced over a 50% increase in the dollar's value, which results in lower import prices and rising export prices of our crops and machines. Figure 2 shows how crops such as grapes, apples and orange juice (all high in labor harvesting costs) are being challenged by foreign competitors. Brazil (Binswanger, et al. 1978) has 100 times less mechanization (tractor-labor ratio) than the U.S. What if they increase this ratio? Will they challenge U.S. markets?

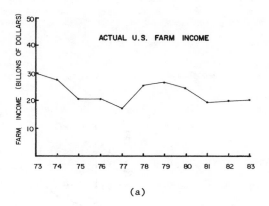

(a)

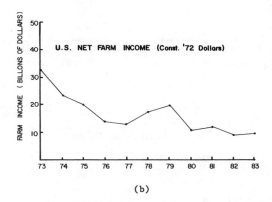

(b)

Fig. 1 U.S. Farm Income. (a) Current Dollars, (b) Constant 1972 Dollars

The farm machinery market is in a similar decline (see Fig. 3) caused mainly
by decreasing farm income and also in part by exchange rates. Retail tractor sales are at a post World War II low; Fig. 3f shows how larger tractors
are continually being imported at a faster rate and small tractors (Fig. 3e)
are almost 100% imported. The only encouraging news in that industry is
that, in terms of dollars, we still continue to increase our farm tractor
exports.

Other reasons for the decline in agricultural machinery sales includes
shifting to less tillage, thus reducing wear and tear on the tractor and extending its life. Also, the costs of overhauling old tractors relative to
purchasing new tractors cause farmers to delay purchases. But farm machinery price increases have exceeded other farm purchase increases as seen in
Fig. 4 thus creating sticker shock at the farm equipment dealership.

The only reason for purchasing farm tractors in the last two years was the
need for U.S. farms to become more productive and efficient relative to
worldwide competition (Fig. 5). The affect of exchange rates on tractor
sales, previously discussed, has been partially offset by applying new production technology to agriculture.

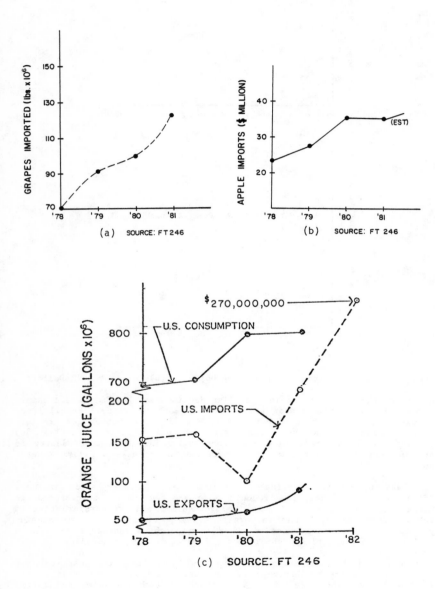

Fig. 2 Agricultural Markets Affected by Imports. (a) Grapes (Chile),
(b) Apples (Canada and New Zealand), (c) Orange Juice (Brazil)

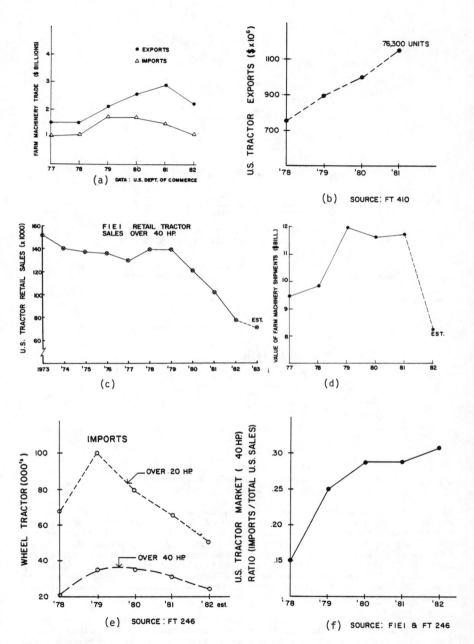

Fig. 3 Farm Equipment Market. (a) Import and Export Dollar Values,
(b) Tractor Exports, (c) U.S. Sales, (d) Total U.S. Market,
(e) Tractor Imports, (f) Import Ratio

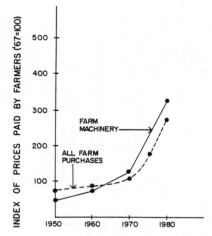

Fig. 4   Increase of Farm Purchase Prices
(Source - Florida Agriculture in the 80's)

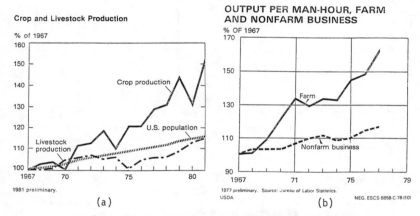

Fig. 5   Agricultural Output and Efficiencies Increases
are the Best in the World

Many other technological advancements on the near horizon will affect agriculture.   The computer is becoming a common tool in reducing farm cost via
simulations.   Computer simulations result in prediction of tighter control
and timing of farming operations.   Biotechnology will drastically change
some production practices and future farm machines.   Battelle (Agriculture
2000, 1983) predicted yield increases up to seven times current averages by
the turn of the century.   Some biotechnologies to watch include tissue culture, CO_2 environments, hormones, nutrient beds, stress tolerant plants,
recombinant DNA, seed coatings and plant environments.

Binswanger et al. (1978) discussed the effect of technical innovation in
U.S. agriculture and concluded there would be a continued shortening of new,
innovative machine adaption time periods.   Adaption of new farm machines has
been price motivated, with the production benefits exceeding implementation
costs.   The trend towards automation of farming will continue with the

application of new technology such as robots, computers, and sensors to reduce production costs.

Low cost computers will make seeing and talking robots a reality. Robot prices will decline in future years in a fashion similar to computer costs as production volumes are scaled up. Ayres and Molburg (1982) states that a robot system costing $80,000 in 1980 might cost $30,000 in 1985. They also estimate payback periods for robots in manufacturing, with different interest rate levels, to be less than 2-years. High inflation rates have recently slowed robot adoption as firms wait for prices to decline.

BENEFITS FROM THE APPLICATION OF ROBOTS
IN U.S. INDUSTRY AND AGRICULTURE

The motivations for using robots differ by application but a survey (Lewis, 1983) demonstrated that reduced labor costs and the elimination of dangerous jobs are high priority reasons, as shown in Table 1.

Table 1. Motivation for Using Robots

1	Reduced Labor Cost
2	Elimination of Dangerous Jobs
3	Increased Output Rate
4	Improved Product Quality
5	Increased Product Flexibility
6	Reduced Materials Waste
7	Reduced Labor Turnover
8	Reduced Capital Cost

The labor question has been controversial. Studies have been done to determine the impact of robots on labor.

The Society of Manufacturing Engineers forecasts that some 20,000 people can expect to lose their jobs during this decade as a result of the introduction of robots to their plants. This number will be far outweighed however, by the 70,000 to 100,000 new jobs expected to be created in the robot support industries. Current automotive labor reductions have been primarily due to the recession, not automation.

Fox (1983) states that the robot industry is still in its infancy. United States sales in 1982 were $195 million, up from $155 million in 1981. Sales of this magnitude are typically generated by 2,000 to 3,000 employees. It is projected that in 1983 United States sales of robots will grow 23%, with a corresponding increase in the number of employees in the robotics industry. By 1990, the robot industry should reach $2 billion. And it should support ten times as many employees as today.

Carnegie-Mellon (Kinnucan, 1983) looked at the labor question and recommended union and management interact cooperatively--rather than as adversaries--for dealing with issues of displacement due to automation. Their report continues to call on government and education to help. They state,

"It would seem that if industry continues its uncommunicative policy, the unions will continue to emphasize setting precedents in order to ensure their survival in an uncertain environment. This type of "gaming" obstructs the type of planning that both unions and management need to do in cooperation with each other to solve real problems and achieve mutual benefits. It is not reasonable to expect firms to be more open with the unions if such disclosures would constrain them in what type of technology they could develop, or how they could use it. Neither is it reasonable to expect unions to be more cooperative with management, and more flexible in their bargaining positions, if such an attitude was considered it would threaten the security of their workers and the long term viability of the unions themselves. The only way for both sides to break out of this bind is for government to change the conditions under which unions and industry talk to each other. In this context, the U.S. may have much to learn from Japan, Germany, and other industrial countries.

"Another of the government's key roles should be to provide incentives which would induce industry to take positive action on upgrading its human resources now. For example, the government could give tax incentives to partially reimburse industry for education and training investments in their employees. It could provide more favorable tax treatment for individuals who undertake formal retraining programs in mid-career. And, of course, it could provide inducements (financial and other) to educational institutions to induce them to redirect their efforts in new areas.

"Education and training are established functions of all levels of government. It is vital that publicly funded education/training programs reflect the emerging--rather than obsolete--needs of industry and society."

Figure 6 examines one of the reasons why robots are displacing labor (their cost per hour). On farms the number of workers is increasing of late and the labor rate has continued to increase to $4/hr. When labor equivalent cost for a robot drops below $5/hr as depicted in the mid-80's in Fig. 6, then robots will be applied on farms. One such crop where robots could be applied is oranges. Industry estimates labor costs and benefits for hand picking oranges to be around $9-10/hr. A continued increase in the number of robots in industry and an application of robots onto the farm has been forecast by many experts (Society of Manufacturing Engineers, 1982) with Japan having the highest usage.

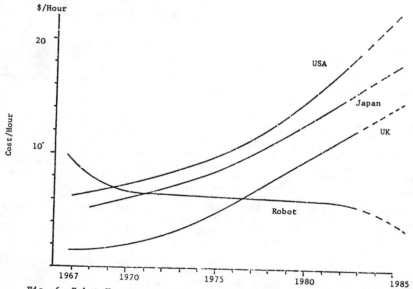

Fig. 6 Robot Versus Human Hourly Labor Cost in UK, USA and Japan

Other benefits could be realized by use of robots in the meat processing industry (again labor intensive). Increased productivity, reduced cost and accidents are key issues in this agriculture industry. Table 2 depicts the current farm accident rate which is projected to decline if and when robots replace farm workers. This is analogous to a declining accident occurrence in industries that use robots.

Table 2. Farm Accidents
==
Summary of Machine Related Accidents[a]

Agricultural Machinery	16.3%
Tractor	8.0%
Truck/Other Equipmemt	15.3%

[a]Data from 1982 Farm Accident Survey Report, National Safety Council. 1900 deaths total in the United States.
==

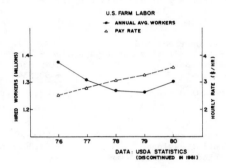

Fig. 7   U.S. Farm Labor

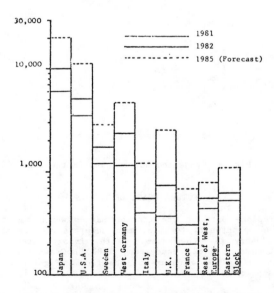

Fig. 8   Forecast of World Robot Population

23

The Delphi Report (Carnegie-Mellon University, 1982) forecasted a decline of over 30% in industrial accidents with robotic usage (directly proportional to robot implementation). Robot usage changes affecting workers (farm or industrial) will be: fewer demeaning jobs; increased training requests; safer, cleaner, less strenous work; less low skill jobs and improved quality of life; more leisure; and higher pay.

Breakeven costs at which robots can be justified may be determined via the following example calculation. Benefits of safety previously mentioned have not been included and would reduce the payback period shown.

An Example Calculation for Potential Savings Using Robots (Lewis, 1983)

Present Process (farm or industrial:

5 operators, overhead 57%
$10.00 per hour @ 1.57 @ 5 @ 1896 hrs/yr = $148,836/yr

Robot Process

2 operators, overhead 57%
$10.00 per hour @ 1.57 @ 2 @ 1896 hrs/yr = $ 59,534/yr
Present labor costs $148,836
Robot labor costs   − 59,534

Total potential savings $ 89,302/yr

Payback

Cost to install robot $125,000
Total potential savings $ 89,302/yr
Payback period − 1 year, 5 months

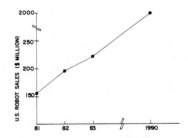

Fig. 9 Projected Robot Sales in the U.S.

FUTURE USAGE OF ROBOTS

Robots might replace part of the tractor market which has been declining. This could be an opportunity for U.S. manufacturers of farm equipment to re-build their industry in a new area. With vision systems (Artley, 1983; David, 1983; Gevarter, 1982; Sanderson, 1983; McGraw-Hill, 1983) becoming reliable, robots will have a future on the farm. Figures 10 and 11 depict some possible applications.

Figure 10a depicts a roving robot checking crop growth and problems. Also, in the same figure, a tissue culture planting robot handles the high value

24

seedling with care. Oranges are harvested quickly in Fig. 10b with the same robot used to fumigate and cultivate (Fig. 10c) or spray (Fig. 10e) when not harvesting. This multiusage reduces the robot's relative labor cost. Other high-labor, high-value crops could become markets for robots such as harvesting melons (Fig. 10d). Having milked dairy cattle, I can foresee potential in reducing labor and controlling the milking process with robots (Fig. 10f).

With time robots will become more intelligent (per programming by humans) and could benefit other agricultural areas such as greenhouse production (Fig. 11), plant breeding, veterinary assistance, tissue culture laboratory work, aqua harvesting, meat packing and doing household chores.

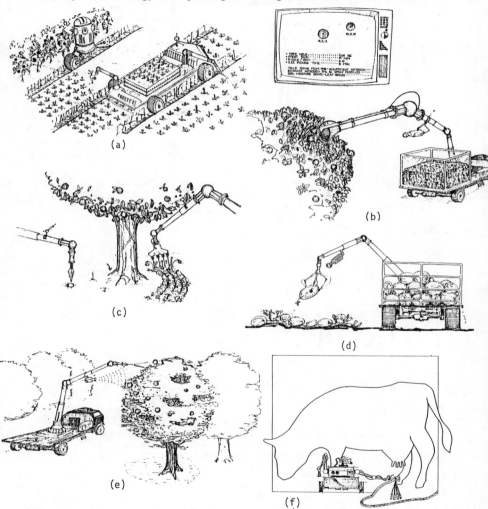

Fig. 10 Future Uses of Robots in Agriculture. (a) Crop Production,
(b) Orange Harvesting, (c) Cultivating, (d) Melon Harvesting,
(e) Spraying, (f) Milking

25

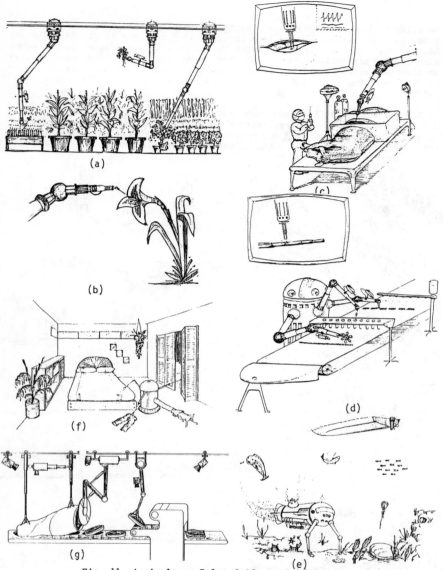

Fig. 11 Agriculture Related Adaptation of Robots.
(a) Greenhouse, (b) Breeding, (c) Veterinary Medicine,
(d) Tissue Culture Lab, (e) Aqua Harvesting,
(f) Home Robot, (g) Meat Processing

SUMMARY

Many ideas have been presented. It can be speculated that a few of these will be on the path of future technological changes. The ones picked will ultimately be determined by the market place.

The excitement for engineers, scientists, and marketing experts is the development of these new machines. Some have been presented, many more are conceivable. David (1983) makes the point that high technology industries and universities have grown due to technology advancements and have created far more jobs than have been lost in manufacturing over the past decade.

In order to achieve a portion of the farm robot alluded to in this paper, demands will be placed on research. Areas short of basic knowledge include physical properties of biological materials, biological sensor technology, spatially adaptive mechanisms and high speed handling processes for new products.

An integrated research effort between industry and adacemia is needed to assure the coexistence of both in this very competitive worldwide marketplace.

Currently, the USDA in their goals for research programs in the 1980's has not yet mentioned robot technology or sensor development integrated with computer control. These devices are needed to keep our farms and industry from declining. Support is needed to develop agricultural engineering robotics research facilities. An example of a robot lab for agriculture costing less than $150,000 is suggested in Fig. 12. The question is, where does the money come from?

REFERENCES

1. Agriculture 2000, 1983, Battelle Press, Columbus, OH.

2. Artley, John W., "Robot Vision-The Future is In Sight," Feb. 1983, Robotics World, pgs. 14-17.

3. Austin, D. W. and Beazley, J. E. "Struggling Industries in Nations Heartland Speed up Automation." Wall Street Journal, April 4, 1983.

4. Ayres, R.U. and Molburg, J.C., "Approaches to Future Labor Impacts of Robotics," Nov. 1982, Dept. of Labor Final Report, B9M22242, Variflex Corp., Box 7771, Pittsburgh, PA.

5. Binswanger, Ruttan, et al. Induced Innovation, 1978, John Hopkins Univ. Press. Baltimore, MD 21218.

6. Brechenridge, P. "Officials Say Hillsborough Schools Stack up Fine against Japanese." March 15, 1983, The Tampa Tribune pgs. 1, 2B.

7. Carnegie-Mellon University. "The Productivity, Human Resources and Societal Implications of Robotic and Advance Information Systems." March 1982.

8. David, E.E., Jr., "By 1990 All Industries Must Be High Tech," April 1983, High Technology, Boston, MA, pg. 65-68.

9. Fox, R. Westinghouse Corp. Orlando, FL., Letter, March 21, 1983.

10. Gevarter, W., "An Overview of Artificial Intelligence and Robotics Vol. II," March 1982, NASA NBSIR 82-2479, Washington, D.C.

11. Hynowitz, C. "Manufacturers Press Automating to Survive, but Results are Mixed." Ibid, April 4, 1983.

12. Ingrassia, P. and Darlin, D. "Cincinnati Mjlacron, a Measley Metal Bender Now is a Robot Maker." Ibid, April 7, 1983.

13. Jamison, D.T. and Law, L.J.   "Farmer Education and Farm Efficiency," 1982. IBID.

14. Kinnucan, P. "Machines That See," April 1983, High Technology, Pgs. 30-36.

15. Krutz, G. W.   "Intelligent Machines for Agriculture in 1990."   Sept. 1983, SAE paper, Milwaukee, WI.

16. Lewis, A.   "Investment Analysis for Robotic Applications."   April 1983.   13th ISIR Robots Proceedings Conf., Chicago, IL, pgs. 4-128 to 4-139.

17. Sanderson, R.J., "A Survey of the Robotics Vision Industry," Feb. 1983, Robotics Worlds, pgs. 22-31.

18. Society of Manufacturing Engineers.   "Industrial Robots" a Delphi Forecast of Markets and Technology.   Pub. by University of Michigan, Ann Arbor, MI.   1982.

19. Taylor, Jim, "From Traction Research to A Crop Production System," Nov. 10, 1982, ASTM Tech Seminar, Akron, OH.

20. Tolley, G. S., Thomas, V. and Wong, C. M.   "Agricultural Price Policy and the Developed Countries, 1982.   IBID.

21. 28th McGraw Hill Survey of Business' Plans for Research and Development Expenditures, 1983-86, May 1983, McGraw Hill Pub. Co.

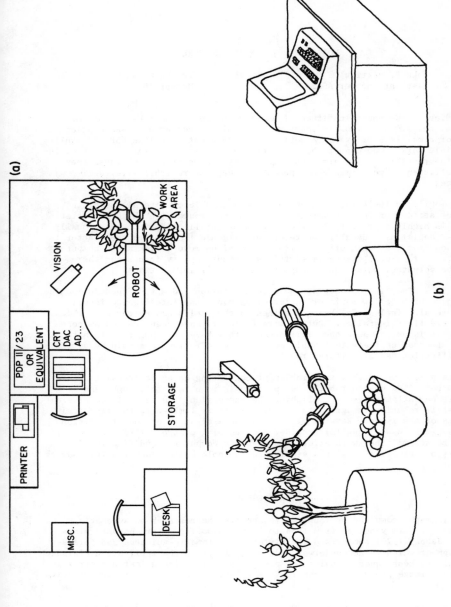

Fig. 12 Proposed Agricultural Engineering Robotics Research Laboratory. (a) Plan View, (b) Experimental Robot Picking Fruit

29

MOBILE ROBOTS IN AGRICULTURE

John B. Gerrish
Associate Member ASAE

Truman C. Surbrook
Member ASAE

According to the Robot Institute of America, a "robot" must be capable of
several tasks. And a mobile robot must be able to get around. No mention is
made of arms, legs or eyes. A mobile robot need not look like C3PO to qualify
for field work. One cannot, however, easily dismiss the suspicion that evolution favored things like arms and legs and eyes--two of each--because they
were efficient ("performance per pound"), cheap and reliable--i.e. well-
designed.

Among mobile agricultural machines which are capable of several tasks, the
tractor stands our for its versatility. The environment in which a tractor
works is highly unpredictable, therefore some intelligence is needed locally
to make decisions. Usually, whatever intelligence is on-board the tractor is
borrowed from its human driver. It is an obvious move to install a computer
on a tractor. Intelligence is worthless without sensory inputs and there
must be actuators which can execute the computer's instructions.

There is a fine line between advanced mechanization and a true robotic machine.
Machines which follow rails or buried wires avoid the intelligence issue;
they are also inflexible, being confined to the fields and patterns in which
they were first installed. Buried wire followers have been adopted as con-
veyors in industry where they fill a need; they have not caught on in agriculture. Because of their inflexible mobility, wire followers should be denied
classification as agricultural robots.

In this paper, we address the problem of converting an agricultural tractor
into a bona fide mobile robot. Our purpose is to stimulate thought and
inspire new attempts at accomplishing the conversion. We leave it to others
to justify robotics in agriculture; our concern is limited to the technical
aspects of the issue. We will (1) present an assessment of computer vision
as a sensory input, then (2) consider the steering control problem and how it
might be accomplished with affordable on-board computers, and finally (3)
discuss our experiences with a simple programmable battery-powered lawn-mowing
tractor.

BACKGROUND

A bibliography compiled by Young (1976) covers the research on automatic
guidance of farm vehicles from 1958 to 1976. Also listed are patents since
1924. Jahns (1976) summarized the history of automatic guidance and forecast
development which is now underway. Jahns anticipated the application of image
recognition teahniques to solve the problem of relocating a lost directrix in
a field. Image processing technology is relatively new; the already extensive

---

*The authors are: J.B. Gerrish and T.C. Surbrook, Associate Professors,
Agricultural Engineering Department, Michigan State University, East Lansing,
MI 48824.

literature is introduced in several texts (e.g. Ballard and Brown, 1982).
Computer processing of an image is a bottleneck in the control system.
If the computer is to be small and cheap, control techniques must be found
which will free the computer to undertake the massive computational task.
A study of how humans economize on mental effort can illuminate the issue
(Hoffmann 1975).

Two devices have appeared on the market which permit driver-less tractor
operation: One was a buried-wire follower and the other is a furrow-
follower. Wire followers lack the flexibility required of a robot; the
furrow followers (or edge followers) suffer from a problem which causes
steering errors-once committed-to be propagated in subsequent passes, and
perhaps amplified. In Japan, a driver-less combine successfully harvested
a rice field (Kanetoh, 1976). We acknowledge the interesting publications
on the "walking machines" (Reibert and Sutherland, 1983) and the work on
vehicle guidance using scene analysis (Norton-Wayne and Guentri, 1981,
Moravec, H.P., 1977, Wirtz and McVay, 1982). Recently the hobbyists have
begun to market small "domestic" mobile robots; best known is the "Hero"
made in Michigan. The appearance of the hobby robot is the surest harbinger
of progress in the development of the mobile robot.

Recent efforts by the MSU research group include an over-the-row apple harvesting machine (Upchurch, 1980 and McMahon, 1983), and a self-guided lawn-
mowing tractor (Surbrook, et al., 1982). The apple harvester features
ultrasonic tracking on tree rows; the lawn mower uses a tactile grass-edge
sensor. The apple harvestor is a hydraulically-driven machine; the lawn
mower is battery powered: both avoid the complication of interfacing
computers with conventional clutches and transmissions.

ROBOT VISION

Three problems have plagued automatic guidance systems: getting lost (i.e.
a lost directrix due to following a false edge or false furrow), constant
hunting (for the ideal trajectory), and error propagation. All of these
problems result from relying on a single sensory input and single closed
loop operation. An improvement is possible if the sensor system is improved
to include a sensor which offers a global view of things, and if possible is
also anticipatory. The edge follower is then supplemented by a sensor which
scans ahead and prevents steering corrections from becoming too drastic.
This is not the same as desensitizing the edge sensor by increasing the
dead band, or increasing the time constant.

New technology provides us with cameras capable of taking in the same view
which the human operator sees and uses for this global input. The question
is whether such cameras can provide sufficient accuracy for putting the tool
on the proper trajectory.

Figure 1 depicts a camera mounted on a field machine and indicates the
important parameters. Roll and pitch are neglected.

Parallel rows or tracks in the field appear on the retina of the camera as
a pencil of lines converging on the horizon at the so-called vanishing point.
Figure 2 shows the view "seen" by a camera for a set of parameters. This
view was computed using a simple geometric transformation; the picture is
generated by an IBM computer.

The heavy line is the desired tool trajectory. x_1 and x_2 can be determined
to $\pm$ pixel (about 0.05 mm) ($\pm$ 1/2 cm for A and $\pm$ 1 cm for B.) Using the
complete information, A and B can be statistically estimated to $\pm$ 1/2 cm$\div$
$\sqrt{128}$ and $\pm$ 1 cm/$\sqrt{128}$ respectively for A and B, given that the line AB on
the field is a perfect line. If AB is fuzzy, e.g. it is a row on onions

and the edge is known only to $\pm$ 10 cm, point B is now accurate to $(10 + 1)/\sqrt{128}$ cm = $\pm$ 1 cm. It is important to realize that the information in·the picture between points A and B is valuable for determining A and B with accuracy.

The lines on the field have the equation

$$X = K - Y \tan \theta \tag{1}$$

K and tan θ can be determined from two measurements on the picture: x_1 and x_2

$$\tan \theta = \frac{x_1 - x_2}{P} \sin \delta - \frac{x_1 + x_2}{2d} \cos \tag{2}$$

and:

$$K = \frac{h(x_1 + x_2)}{2d \sin \delta} \tag{3}$$

where θ is the heading error with respect to the rows, (θ positive if heading to the right of the vanishing point)

d = the distance from the pinhole to the retina of a pinhole camera.

p = the retina height

h = the camera height above the ground plane.

δ = the tilt angle measured down from horizontal.

X,Y = field coordinates with respect to origin at intersection of camera axis with the field plane.

x_1, x_2 = coordinates on the retina; x_1 is at the botton edge of the retina, x_2 is at the top edge.

If the tool is located s meters behind the camera and w meters to the right of the camera axis, then Y = - (h cot δ + s) $\tag{4}$
and:

$$X = h \frac{(x_1 + x_2)}{2d \sin \delta} + \left(\frac{h}{\tan \delta} + s\right)\left\{\frac{x_1 - x_2}{P} \sin \delta - \frac{x_1 + x_2}{2d} \cos \delta\right\} \tag{5}$$

We define an offset error ϵ which is positive if the tool is too far to the right of the desired trajectory.

$$\epsilon = w - \left(\frac{x_1 + x_2}{2d} (h \sin \delta - s \cos \delta) + \frac{x_1 - x_2}{p} (h \cos\delta + s \sin \delta)\right] \tag{6}$$

In general, for a row to the right of the machine, $x_2 < x_1$. ϵ is the "error signal" upon which steering control will be primarily based; θ provides the global input - the directrix. Knowing θ, we will dare to interrupt closed loop operation since there is a reduced risk of getting lost. θ will also limit the correction rate (preventing error propagation) and may even become the preferred guidance input should a rapid $d\epsilon/dY$ cast doubt on the validity of ϵ.

The merit in the vision system is that both local and global inputs can perhaps be obtained. An error analysis on the error signals (θ and ϵ) is instructive.

The absolute error in the heading error measurement is

$$\Delta\theta = \left|\frac{\partial\theta}{\partial x_1} \Delta x_1\right| + \left|\frac{\partial\theta}{\partial x_2} \Delta x_2\right| + \left|\frac{\partial\theta}{\partial\delta} \Delta\delta\right| \qquad (7)$$

The maximum absolute error in the offset error ε is

$$\Delta\varepsilon = \left|\frac{\partial\varepsilon}{\partial h} \Delta h\right| + \left|\frac{\partial\varepsilon}{\partial s} \Delta s\right| + \left|\frac{\partial\varepsilon}{\partial w} \Delta w\right| + \left|\frac{\partial\varepsilon}{\partial\delta} \Delta\delta\right| + \left|\frac{\partial\varepsilon}{\partial x_1} \Delta x_1\right| + \left|\frac{\partial\varepsilon}{\partial x_2} \Delta x_2\right| \qquad (8)$$

where Δ's indicate the maximum uncertainty in a measurement, $\Delta\varepsilon$ being the total of all the individual contrictions to uncertainty. d and p are presumed known to great accuracy; therefore, their error terms and neglected.

For small heading erros, $\theta \simeq \tan\theta$. Table 1 lists the partial derivatives. Note that W, S and h have no effect on the accuracy of θ. To get the most accurate heading error signal, the camera must be located in line with the tool (w = 0) and the desired trajectory ($x_1 = x_2 = 0$, when on-course). Notice that the offset error is also improved by this location; the offset error will be immune to vertical vibrations or uncertainty of the camera position (h) as well as variations in pitch (δ). Not much can be done about Δw; its full effect is felt on the error signal. But if w is zero, Δw can probably be limited to less than 10 mm. Some typical values are:

h = 1.5 m, d = 6.5 mm, p = 6.0 mm, Δh = 0.01 m, Δs = 0.01 m, $\Delta\delta$ = 1° (0.017 radians), $\Delta x_1 = \Delta x_2 = 0.05$ mm

Table 1. Partial derivatives which appear in Equations 7 and 8.
===

$$\frac{\partial\theta}{\partial x_1} = \frac{\sin\delta}{p} - \frac{\cos\delta}{2d}$$

$$\frac{\partial\theta}{\partial x_2} = \frac{-\sin\delta}{p} - \frac{\cos\delta}{2d}$$

$$\frac{\partial\theta}{\partial\delta} = \frac{x_1 - x_2}{p} \cos\delta + \frac{x_1 + x_2}{2d} \sin\delta$$

$$\frac{\partial\varepsilon}{\partial h} = \frac{-x_1 + x_2}{2d} - \frac{x_1 - x_2}{p} \cos\delta$$

$$\frac{\partial\varepsilon}{\partial s} = \frac{x_1 + x_2}{2d} \cos\delta - \frac{x_1 - x_2}{p} \sin\delta$$

$$\frac{\partial\varepsilon}{\partial w} = 1$$

$$\frac{\partial\varepsilon}{\partial\delta} = \frac{-h}{2d}(x_1 + x_2)\cos\delta + \frac{h}{p}(x_1 - x_2)\sin\delta \; \frac{-s}{2d}(x_1 + x_2)\sin\delta - \frac{s}{p}(x_1 - x_2) \bullet \cos\delta$$

$$\frac{\partial\varepsilon}{\partial x_1} = \frac{-h}{2d}\sin\delta - \frac{h}{p}\cos\delta + \frac{s}{2d}\cos\delta - \frac{s}{p}\sin\delta$$

$$\frac{\partial\varepsilon}{\partial x_2} = \frac{-h}{2d}\sin\delta + \frac{h}{p}\cos\delta + \frac{s}{2d}\cos\delta + \frac{s}{p}\sin\delta$$

===

33

Note that if $x_1 = x_2 = 0$ then both heading and offset errors are insensitive to errors in δ or the variations in δ caused by pitch. Moreover ε will be insensitive to errors in h (caused by vertical vibrations) and δ (which should give no problem). If $x_1 = x_2 = 0$, then only the Δx_1 and Δx_2 terms contribute to the error in the offset error, ε. It can be shown that a minimum $\Delta\varepsilon$ occurs when

$$s = -h \cot (\delta + \arctan p/2d) \qquad (9)$$

The minimum cannot be determined by the usual method of calculus since $\Delta\varepsilon$ is a sum of absolute values. Substituting $\hat{s}$ for s leads to

$$\left.\frac{\partial\varepsilon}{\partial x_2}\right|_{s=\hat{s}} = 0 \qquad (10)$$

meaning that the determination of ε is insensitive to inaccuracies in measuring x_2 (at the top of the picture). This is because the tool is in the image and ε can be seen directly be measuring the distance between the tool (on the vertical axis of the picture) and x_1, the intercept of the desired trjactory with the bottom edge of the picture.

$$\Delta\varepsilon \text{ min} = \left|\frac{\partial\varepsilon}{\partial x_1}\Delta x_1\right|_{s=\hat{s}} = \left|\frac{-2h\,\Delta x_1}{2d\,\sin\delta + p\,\cos\delta}\right| \qquad (11)$$

The surprise here is that there is a best location for the camera and it is not on the "hood ornament" but rather, in line with the tool, above and behind it such that the tool is centered at the bottom edge of the image. Equation (9) vindicates all mothers-in-law who have been found guilty of back-seat-driving.

To get the minimum error in the offset error, set:

$$\frac{\partial(\Delta\varepsilon \text{ min})}{\partial\delta} = 0 \qquad (12)$$

This leads to:

$$\tan\delta = 2d/p \qquad (13)$$

The heading error uncertainty when on-course ($x_1 = x_2 = 0$) is:

$$\Delta\theta = \left|\frac{\partial\theta}{\partial x_1}\Delta x_1\right| + \left|\frac{\partial\theta}{\partial x_2}\Delta x_2\right| = \left|\frac{\sin\delta}{p} - \frac{\cos\delta}{2d}\,\Delta x_1\right| + \left|\frac{-\sin\delta}{p} - \frac{\cos\delta}{2d}\Delta x_2\right| \qquad (14)$$

The Δx_1 term goes through a discontinuity which is a minimum at

$$\tan\delta = p/2d \qquad (15)$$

and:

$$\Delta\theta\text{min} = \Delta x_2 \frac{2}{\sqrt{p^2 + 4d^2}} \qquad (16)$$

(Δx_1 has no effect on $\Delta\theta$ min).

Thus we have a dilemma. To get an accurate heading error, $\delta = \arctan p/2d$ (**24.8°** in our case with the horizon just in view at the top edge of the image). To get an accurate offset error, $\delta = \arctan 2d/P$ (**65.2°** in our case, with the tool located directly beneath the camera pinhole).

34

One solution would be to increase p relative to d. This would permit "seeing" the tool and the horizon simultaneously. Increasing d helps reduce offset error uncertainty but does nothing for the heading error uncertainty. If p is increased, the number of pixels should be increased. If the ratio of p/2d becomes], wide angle lenses and their inherent curvature may distort the image which an ideal pinhole camera would create. A typical digitizing camera has p = 6 mm and d = 6.5 mm with a pixel measuring 0.05 mm on a side (128 x 128).

It is reassuring to note that even in cases far from optimum (e.g. S = 30°, S = 3 m behind the camera, h = 1.5 m, x_1 = 3 mm, x_2 = 1 mm), $\Delta\epsilon$ and $\Delta\theta$ are still tolerable ($\Delta\epsilon$ = 5.94 cm excluding the Δw term; $\Delta\theta$ = 0.0084 radians or 0.5 degrees). If everything were optimal (δ = 45°, s = 0, p = 2d = 13 mm), then $\Delta\epsilon$ = 8.2mm and $\Delta\theta$ = 0.0054 radians or 0.31 degrees.

STEERING CONTROL STRATEGY

The minimum-error locations of the camera and a rigidly-mounted tool are shown in figure 3 for a tilt angle of 30°, d = 6.5 mm, p = 6 mm, and h = 1.5m. For these conditions, s = 1.06 m and $\Delta\epsilon$ = 12.8 mm. The tractor is 2 WD and is massive enough with respect to the tool that the center of the rear axle is the center of rotation for the tractor-tool combination.

The problem, of course, is that a steering correction applied in response to a heading error, causes the offset error to change. This interaction can be analyzed, or it can be treated by a method of anticipated trajectory as is the usual case in accomplishing robot arm tracking. Following robotic techniques, the tractor is permitted to operate "in open-loop mode" (i.e. without feedback) for a brief time during which corrections are made. The duration of the open-loop episodes depends on how confident one is in both dead-reckoning and the ability to re-register the row should it become lost. Obviously

$$T_{ol} \leq \min \frac{R}{2 \text{ v sin } \theta} > \frac{\epsilon \text{ max}}{v \sin \theta} \qquad (17)$$

where R is the row spacing
v is the forward velocity
ϵ_{max} is the maximum tolerable offset error.

This method of steering might be called pseudo closed loop. It mimics the way our man-machine interface operates when we drive an automobile. Automobile drivers operate in three modes: 1) a closed-loop mode when constant attention is required as in dangerous situations, 2) intermittent control of lateral displacement from the course and, 3) a predictive method in which the future position on the roadway is calculated based on current heading and offset. (Weir and McRuen, not seen. Referenced in Sweatman and Joubert, 1976). Whereas the control of a tractor or combine differs somewhat, the human operator unburdens his/her brain by reverting to an open loop strategy as often as possible. In order to free an on-board control computer for the slow image acquisition and processing techniques, a similar strategy is in order. Equation (17) suggests that only if θ is nearly zero does one dare to begin an open-loop episode. Humans perform steering corrections intermittently also to conserve steering muscle energy.

The method for correcting a heading error is simply to cause the front wheels to move to a predetermined limit angle, α_{max}, dwell there, and return the wheels to a centered position, α = 0. To obtain small corrections, α_{max} can be made small, but the inevitable "play" in the steering linkage limits this option. The dwell time can also be reduced to a limit imposed by "play". The patterns for α which would correct a heading error and an offset error are:

35

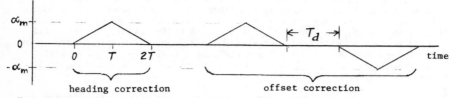

heading correction   offset correction

"Play" is neglected in these patterns. The slope of the ramp is very likely a fixed value determined by the steering actuator. If the minimum increment heading correction is:

$$T = \frac{1}{\alpha'} \; arccos \; exp \; (-\frac{\theta_1}{2v} \; \alpha'b) \tag{18}$$

where $\alpha' = d\alpha/dt$ for the steering actuation and b is the wheelbase.

Obviously a heading correction would not be started unless $|\theta|$ exceeded θ_1.

The tool is moved laterally during a heading correction. If the initial heading angle θ_0 is small and θ_1 is also small, then $\sin \theta_1 \approx \theta_1$ and

$$\varepsilon - \varepsilon_0 \approx a \; (\theta_1) = -a \; \frac{v2}{b\alpha'} \; \ln \cos \alpha' \; T \tag{19}$$

where a is the distance the tool is ahead of the rear axle of the tractor.

In an offset correction, the two heading corrections give compensating changes in ε. The net effect is due to the dwell time. If θ_1 is the unit correction, then

$$\varepsilon_1 = T_d v \sin \theta, \; or \; simply \; T_d v \theta_1 \tag{20}$$

where ε_1 is the correction in ε accomplished in a dwell time T_d

If the camera is positioned such that the tool can be seen at the bottom of the picture, then ε predicts the value of x_1 after a correction.

The value of x_2 after a heading correction will be found from equation (1). These values will help to re-register the image and remain indexed on the correct row after an episode of open-loop operation.

A typical value for α' is 30° per second or about 0.5 radian/second with $\Delta\theta$ on the order of 1/2°, let $\theta_1 = 1.5$ degrees or 0.025 radians. Let the wheelbase be 2 m and v = 4 m/s (9 mph). $T \approx 150$ ms and a heading correction in open-loop mode would free the computer for 300 ms to engage in image processing. We estimate that this is adequate time for a 68000 computer with a 32 K core memory to process the data from a 128 x 128 pixel image, find rows and compute errors. For a tool 1.2 m ahead of the rear axle, each 1.5° heading change will cause an offset error change of 30 mm (in the same direction if the tool is ahead of the rear axle. If an offset correction is required the system would be in a visual blackout for 4T plus the dwell time. Vision could be confirmed after 2T; then T_d of 300 ms would give a minimum ε_1 of 30 mm.

It is important to realize that not every error situation demands immediate action. Table 2 illustrates the point that perhaps only half the errors require action.

EXPERIENCE WITH A PROTO-ROBOT

A battery powered lawnmower was fitted with a linear
actuator (to the steering link), a grass-edge sensor (two sail-switches), a
wheel position indicator (a potentiometer) and an RCA 1802 microprocessor
with 2 kilobutes of memory. One drive wheel had magnets which caused a reed
switch to operate and give an indication of the distance travelled. Other
details may be found in Squires (1981).

On dry grass, the steering actuator turned the wheels through 40 angular
degrees in one second. The wheelbase was 1.0m and the speed was only 0.2
m/s. A distance pulse occurred every 0.3 meters or about every 1.8 seconds.
Any open loop operations had to be accomplished within about 1.8 seconds in
order that the next distance pulse could be received. Each heading correction
was about 1.1° or 0.019 radians and took about 1.7 seconds. During a
correction, no input signal could be processed. Even with nearly one second
blackouts, the controller performed well and the edge was not lost. The
very low speed meant that only 0.2 m was covered during the blackout. With
perhaps 5° deadband (including play) the maximum offset error which might
occur is 17 mm. The sensor deadband was larger than 17 mm. So the
intermittent control was as good as closed loop control for our humble needs.

We wish to report early indications that edges can be discerned in black-
and-white photos of some field scenes which might provide guidance for a
future mobile robot. We need to establish the accuracy with which x_1 and
x_2 can be determined statistically when the image is fuzzy due to vibrations
and the rows are ill defined. Except for this last concern, it appears
that a mobile robot is capable of certain agricultural field operations
and the robot can be built now for about 115 percent of the cost of the
tractor.

CONCLUSIONS

1. There is a best location for the camera with respect to the tool: it
    should be above and behind the tool such that the tool is seen in the
    middle of the lower edge of the uninverted image.

2. It is desirable to know the heading in order to avoid attending to
    self-correcting offset errors.

3. With current technology it is possible to measure tool offset errors
    to within $\pm$ 11 mm plus an error which we have not determined in finding
    the desired trajectory is a noisy image.

4. The heading error can be measured to $\pm$ 0.3 angular degrees. Plus an
    error (which we have not determined) caused by ill-defined rows.

5. A camera with a 90° cone angle tilted at 45° would further minimize
    measurement errors.

6. Steering algorithms which put the robot temporarily in a dead-reckoning
    or open-loop mode will free a computer for adequate time to process
    image information. It appears that a velocity of 4 m/s (9mph) matches
    a path deadband of 60 mm and a heading deadband of 3 °. 300 ms is
    available for processing an image; this amounts to one image every 1.3
    m.

7. Intermittent heading corrections were satisfactory for controlling a
    lawn-mowing driverless tractor.

REFERENCES

1. Ballard, D. and C. Brown. 1982. Computer Vision, Prentice Hall.

2. Hoffmann, E.R. 1975/76. Human control of road vehicles. Vehicle System Dynamics 5:105-126.

3. Jahns, G. 1976. Possibilities for producing course signals for the automatic steering of farm vehicles (in Young, 1976).

4. Kanetoh, I. 1976. Driverless Combine Harvester. Grain and Forage Harvesting Conference Proceedings. ASAE.

5. Moravec, H.P. 1977. Towards automatic visual obstacle avoidance. Proc. 5th Int. Joint Conference on Artifical Intelligence, Cambridge, MA p. 584 ff.

6. Norton-Wayne, L. and D. Guentri. 1981. Vehicle Guidance by Automated Scene Analysis. Proc. Automated Guided Vehicle Systems. Stratford-upon Avon, U.K. pp. 129-136.

7. Raibert, M.H. and I.E. Sutherland. 1983. Machines that walk. Sci. Am 248:44-53.

8. Squires, R.E. 1981. An automatic steering controller for a robotic lawn tractor. M.S. Thesis. Michigan State University.

9. Surbrook, T.C., J.B. Gerrish, R.E. Squires and M.A. Schanblatt. 1982. An automatic steering controller for a robotic lawn tractor. ASAE Paper No. 82-3039.

10. Sweatman, P.F. and P.N. Joubert. 1976. Automobile Directional characteristics and Driver Steering Performance. Vehicle System Dynamics 5:155-170.

11. Upchurch, B.L., B.R. Tennes and T.C. Surbrook. 1980. Development of a microprocessor-based steering controller for over-the-row apple harvester. ASAE Paper No. 80-1556.

12. Wirtz, M.J. and E.S. McVey. 1982. A pattern recognition system for measuring vehicle-to-road-edge distances on a moving vehicle. Proc. 14th S.E. Symposium on Systems Theory (IEEE).

13. Young, R.E. 1976. Automatic guidance of farm vehicles: A monograph. Agricultural Experiement Station, Auburn University, Auburn, AL.

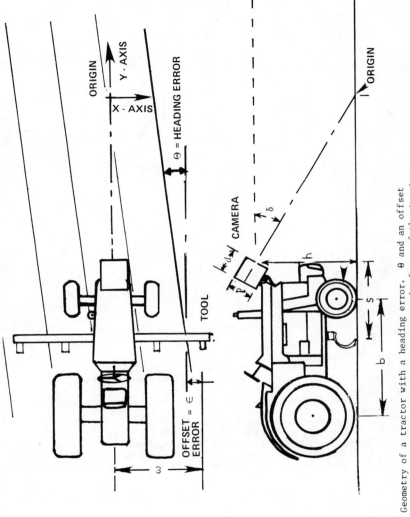

Figure 1. Geometry of a tractor with a heading error. θ and an offset error ε. A downward-looking camera at the front of the tractor creates an image on which guidance will be based.

39

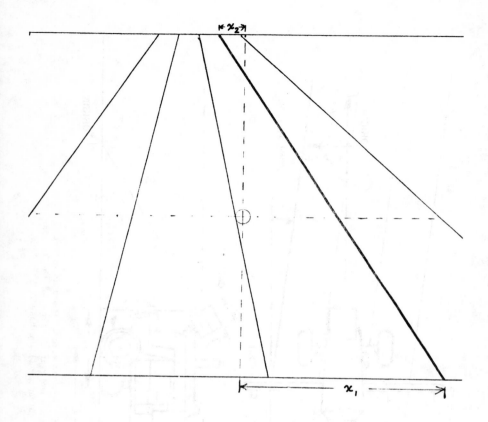

Figure 2. A computer-generated image of the field of parallel "rows", 0.75m The camera is mounted 1.5 m above the ground; tilt is 30°; the retina is 6.5 mm behind the pinhole. Apart as seen by the camera in Figure 1, the tool is 1.5 m outboard on the right, 1.2 m behind the camera, then the offset error will be +200 ± 55.6 mm, i.e. it is to the right of the planned trajectory. The heading error is 6.5 ± 0.48 degrees. Action is called-for since the positive heading error is in the wrong direction for correction of the offset error. The retina is 6 mm high. A pixel is 50 mm on a side for a 128 x 128 array. x_1 and x_2 permit calculations of heading and offset errors.

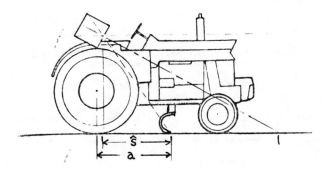

Figure 3.   A better location for the camera is behind the tool with the
tool in the field of view, at the bottom edge. Camera, tool and
trajectory should line up in the same vertical plane. For the
conditions of Figure 2, heading error = 6.5 $\pm$ 0.48 degrees and
offset error = 200 $\pm$ 12.8 mm.

HEADING ERROR

	$\theta < -1.5°$	$-1.5° \le \theta < 1.5°$	$1.5° < \theta$
30mm < ε	-	offset correction required	heading correction required
-30 < ε < 30mm	-	-	-
ε < -30mm	heading correction required	offset correction required	-

OFFSET - ERROR

Figure 4.   Error classification scheme. Not all errors require action.

41

ANIMAL POSITIONING, MANIPULATION AND RESTRAINT
FOR A SHEEP SHEARING ROBOT

S.J. Key and D. Elford

This paper outlines research into an alternative method of harvesting wool from sheep; automated mechanical shearing. The method of animal manipulation is examined. The development of shearing positions; devising restraining methods and the construction of an experimental manipulator is explained. State variable logic is used to monitor and control the manipulator with 40 interacting mechanisms from a sequencing computer program. Hydraulic check valves and cross port relief valves have been used to apply firm yet comfortable pressure to the sheep.

The aim of our research is to devise a means of shearing sheep automatically. The main motivation for doing this lies with the persistent decline in the wool industry's terms of trade in the last two decades. Wool price increases have been limited by competition from synthetic fibres; yet the cost of production has been escalating at a higher rate. The cost of shearing sheep, expressed as a proportion of the wool sold has, on average, doubled in the last 20 years. Secondly shearing is an arduous task, often in isolated and unpleasant conditions as a result there is risk of a shearer shortage in the future.

The Australian Wool Corporation considers automated shearing as a possible future alternative wool harvesting method. The Corporation is supporting this project as part of a national research programme to explore both novel wool harvesting techniques such as biological

defleecing, and improvements to conventional technology, such as long life cutters.

An experimental robot has been built and used for trials during the last three and a half years. Over 200 sheep have been shorn by the machine (though not completely) yet only a few cuts have occurred. This extremely low injury rate results from the use of sensors mounted on the shearing cutter which allow the computer controlling the robot to keep the cutter moving just above the sheep skin.

Figure 1 illustrates how the robot has been used. The sheep is supported on a manipulator and is restrained firmly by the legs and head. In this way the sheep feels that it cannot escape and yet is fairly comfortable and will remain quite still for long periods of time if required. The shape of the sheep changes with breathing and with

The authors are S.J. KEY, Research Engineer and D. ELFORD Consultant, Mechanical Engineering Department, University of Western Australia.

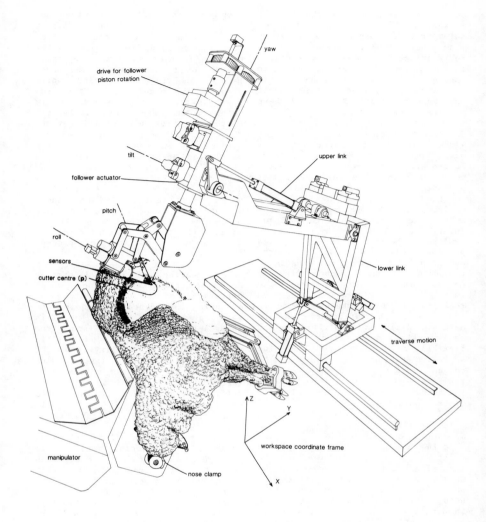

yaw

drive for follower
piston rotation

tilt

follower actuator

pitch

roll

sensors

cutter centre (p)

upper link

lower link

traverse motion

Z

Y

workspace coordinate frame

X

manipulator

nose clamp

Fig 1. Arrangement of Experimental Shearing Robot.
Details of Hydraulic and Electrical Connectors Have
Been Omitted.

43

occasional movements of the animal; further, the shape of any particular animal differs to a surprising extent from other animals. However, by registering the sheep in the cradle in a repeatable manner, one can obtain a reasonably predictable relationship between the basic physical characteristics of the sheep and its surface shape.

The robot consists of a series of links moved by hydraulic actuators using proportional analogue servo controls. The two actuators which move the main links have additional pressure feedback compensation. The robot is controlled by a single minicomputer through a conventional multiple channel digital to analogue interface. The minicomputer is a Hewlett Packard 21MX-E which presently has 256k bytes of memory, and graphics display. The sensors on the robot transmit signals directly to the computer through an analogue to digital interface. This design reflects the experimental nature of the robot; the structure of the hardware has been simplified as far as possible so that maximum flexibility can be retained for software development. However, the mechanical arrangement of the robot reflects software constraints such as the need for a real time inverse linkage solution.

A fundamental part of the software is a map of the sheep's skin shape. The shearing movements are programmed in relation to fixed points on the map. The map is only approximate but it enables the robot to approach the sheep until the cutter is close enough for the sensors mounted on it to detect the skin. Once the cutter has 'landed' on the sheep the program allows it to follow a path defined in the model surface, even though the actual path of the cutter lies above or below, on the actual sheep surface. The surface model also helps the program 'anticipate' the sheep surface ahead of the cutter.

Unfortunately, sheep differ to such an extent that a single surface model will not suffice for all sheep. If the sheep surface deviates too far

from the computer's model, then the robot gets 'lost' and can no longer find a path over the actual surface which corresponds to a path on the model. To overcome this, the robot has to adapt its model of the sheep before shearing, and also while shearing is occurring. First the sheep is measured before shearing to determine its weight, overall length, width in three places and height. These measurements are used to predict a surface model for the particular sheep on the basis of a statistical correlation obtained by measuring a large number of sheep. However, there are surface variations which cannot be predicted from these measurements, and further, the sheep moves from time to time. So the model still has to be further adapted during the actual shearing process (Trevelyan et al 1982).

After shearing, the robot determines how the predicted surface model differs from the actual sheep by measuring the adaptation required on the predicted model. This information is used to improve the prediction on the next sheep. This self learning capability and the ability to work with live, moving animals makes this robot unique. The fact that the robot can do this while still working at about the same speed as a human shearer places it well apart from other industrial robots.

A further requirement for automated shearing is positioning, manipulation and restraint of the sheep, the subject of this paper, so that all parts of the sheep come within the dextrous shearing range of the robot. Sheep positioning, manipulation and restraint will be referred to collectively as sheep manipulation.

An experimental sheep manipulator has been constructed for use with the robot. Unlike the robot, where much of the complexity can be incorporated as 'software' the manipulator relies on both complex carefully shaped mechanical components to position and restrain the sheep and a flexible, user programmable sequencing

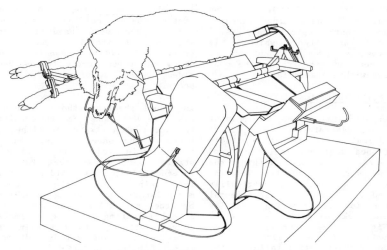

Fig 2. Simplified Illustration of Manipulator

controller to manipulate sheep. The manipulator can be considered to be a a robot in its own right. Whereas the control system for the shearing robot must learn and adjust to its environment as it operates, the manipulator controller relies on the execution of a multitude of predictable events allowing a transition of the arrangement of the mechanisms to occur. These mechanisms possess a self adjusting capability. Figure 2 presents a simplified illustration of the manipulator.

Looking to the future, the AWC expects to decide whether or not to proceed towards the development of automated shearing by late 1983. It is now examining whether the technology will fulfil the needs of the market, whether it is practicable and economically acceptable, and the medium and longer term ramifications for the wool industry. In the event of an affirmative decision, several years of engineering design and development will be required before a field test prototype will emerge. A commercial decision will depend on this further assessment.

If one were to speculate about the likely form of these automatic shearing machines, then they would probably be mobile; each comprising of several robots, and because of their expense they would be required to operate more than one shift per day. Of the other associated wool harvesting tasks, whether automatic wool sorting is acceptable is highly uncertain; however, automatic sheep races, catchers, loaders, measuring and weighing equipment will be more easily incorporated. Figure 3 represents a simplified impression of a likely arrangement. With the present distribution of sheep in Australia, it will never be economic to shear all sheep by machines. Manual shearers will continue to be needed for the foreseeable future. The purpose of building automated shearing machinery is to ensure that there will be adequate shearing capacity at a viable cost to enable the wool industry to continue to make its present major contribution to the Australian economy. Wool exports contribute nearly $A2 billion of export income which is approximately $A130 per head of population.

The present research team consists of seven full time research staff, and six support staff with a budget for the present financial year of approximately $A550,000. The national wool harvesting research programme costs approximately $A1.7 million annually which is approximately seven per cent of the funds spent on research in the wool industry within Australia.

1 Sheep Manipulation

The experimental sheep manipulator positions, restrains and manipulates the animal to present suitably conditioned regions (or patches) of the sheep's surface within the dextrous workspace of the shearing robot. The robot was originallly designed for experimental investigation of shearing radially from the backbone to the belly line. (ref Fig 1). The dextrous workspace is confined to a quadrant of a cylinder with an axis parallel to the x workspace co-ordinate and bounded by vertical and horizontal planes. The limited workspace of the robot has significantly constrained the manipulator design.

1.1 Sheep Positioning: A number of static sheep positions have been defined and were used in early shearing feasibility experiments (ref. Leslie 1980). These positions allow the robot access to the back, sides, belly, neck and rump of the sheep.

In manual shearing, the shearer uses his free hand and legs to condition the loose skin stretching either the body of the sheep or the skin to remove skin folds and present a stable smooth surface for the intended blow; thus avoiding cutting the skin. In the automated manipulator, conditioning is achieved by the manner of supporting the animal resulting in an approximately convex surface with the skin well stretched over a firm body. Of equal importance to the suitability of the position is consideration of the blow patterns that the robot may make in shearing the patch. Allowance must be made for initial entry of the cutter onto the skin. Manual

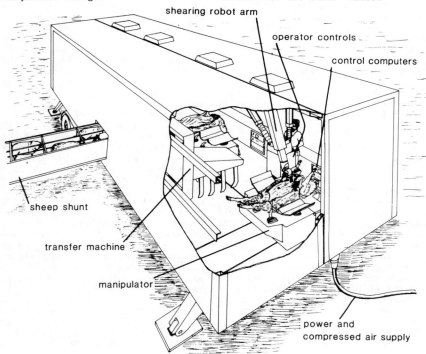

Fig 3. Schematic Automated Mechanical Shearing Plant Arrangement

shearers utilise the area beneath the armpit on the brisket (or chest); an area relatively free from wool. As the drag force on the cutter increases markedly with the distance from the skin, entry through the fleece is unacceptable. The blow pattern must also allow shorn wool to flow away from the area where cutting is taking place. This prevents jamming of wool in the cutter mechanism or re-cutting of wool by it falling in front of the cutter on a subsequent pass. Support of the shorn wool is also important. The fleece tends to hang together and if unsupported will pull up a skin fold that can easily be shorn off causing injury to the sheep.

The four selected and refined shearing positions in present use are illustrated in Fig 4.

1.2 **Sheep Restraint:** Considerable research effort has been required to develop restraint devices that comfortably and reliably hold sheep in the positions identified. Manual shearers can react and anticipate sheep movement and thereby adjust

```
Load sheep
    |
Shear belly left
    |
Rotate sheep
    |
Shear belly right
    |
Transfer to side right position
    |
Shear front leg side and rear leg patches
    |
Roll into back position and rotate
    |
Shear back patch
    |
Rotate and shear rump
    |
Transfer into side left position
    |
Shear front leg side and rear leg patches
    |
Roll into back position and complete as for the right back and rump patches
    |
Unload sheep
```

Fig 5. Manipulation Strategy

their grip, whereas automated restraints must provide an adequate self adjusting grip, yet not injure the animal. Compactness is also of extreme importance allowing maximum access to the wool and easy storage when not in use.

Sheep, like humans, possess several pressure points which cause temporary immobilisation of both the front and rear limbs in the extended position by firm application of a localised grip. These points are utilised for restraint in the belly position. A further useful idiosyncrasy exploited for the initial loading and capture of the sheep in the restraint is the effect of rapid inversion. Natural momentary immobilisation of the muscular responses results, allowing sufficient time for the application of primary leg and head restraints. (ref Ewbank 1968).

Research has shown that a comfortable yet firm grip is needed for reliable restraint. This is achieved by the design of the actuation circuits. The applied force is limited by

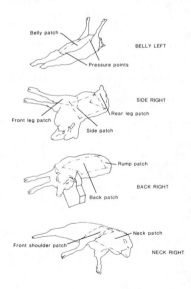

Fig 4. Shearing Positions

pressure regulation. However, any attempt by the animal to reverse the applied force is resisted by the non-return nature of the circuit. Loss of applied force by relaxation of the sheep is overcome by continual application of the valve.

1.3 Sheep Manipulation: Refers to the ordering of the sheep positions and application of restraints to convey the sheep on the device so that it may be completely shorn. Ordering of the sheep positions requires allowance for safe loading and unloading of the sheep, a progressive flow of wool between each position and the minimisation of entry through the fleece.

Figure 5 illustrates the present manipulation strategy. The sheep is first secured in the belly up position, located axially against shoulder gauges and restrained by 'pressure point' restraints in the stifle and armpits. A nose clamp is manually applied leaving the sheep ready for belly shearing. The side shearing positions require the restraint of the animal's legs by secondary restraint mechanisms capable of being rotated with the sheep to both right and left positions. Foreleg clamps are positioned to automatically gather up the two foreleg hocks and stretch them to avoid free leg movement. The rear legs are gathered up by being first raised to a suitable height to be automatically inserted into a dual clamping device capable of applying a controlled combined clamping torque and stretching force. The previously applied foreleg, and rear leg restraints are then withdrawn clear of the work area.

Side panels are now deployed to support the sheep in subsequent side shearing positions. To safely rotate the sheep to the desired side, the secondary leg restraints are locked by means of a hydraulically activated key to the frame on the side which is to carry the sheep. Likewise the relevant neck rest and nose clamp cable are locked to rotate as the one assembly. The weight of the sheep is carried by an interleaved backrest

panel which is subsequently withdrawn to permit access for shearing the animal's back.

Transfer to the other side position requires rotation back through the belly up position. The relevant transfer locks are activated to permit the opposite side panel to carry the neck rest and secondary leg clamps

2 Manipulator Control System

Control of the manipulation function is achieved by two interacting systems. A computer sequencer, monitor and driver package run on the sheep shearing control computer and hydraulic or pneumatic control circuits housed on the manipulator. These systems being interfaced by conventional digital multiplexing and switching electronics. Figure 6 illustrates the functions performed by each of these systems.

2.1 Software Package: The computer software package allows the operator to program each motion of the manipulator mechanisms in a simple coded array called a transition file. Each transition file may contain sufficient operations to manipulate the sheep from one stable position to the next. Files may be nested up to ten deep and may be recursive. The transition file, therefore, becomes a macro command that may be called from the shearing control software (Key 1983).

As each mechanism of the manipulator performs a specified task on the sheep, it may interfere with another mechanism that may operate on a similar area of the sheep. This is especially the case around the head and legs of the animal. Use of conventional interlocks to prevent accidental collision of the mechanisms would greatly complicate the manipulator. Wiring routed through tortuous paths, hard-wired logic and numerous limit switches would be required. A control philosophy has been adopted, therefore, to allow the condition or state of each actuator to be uniquely defined at any instant. The programmer then specifies within the allowable or admissible states of the other mechanisms that must be achieved before the operation can proceed. These contingent states are called preconditions.

A current research consideration is to use linked lists or influence trees to test for admissible state requirements before allowing the proposed movement to commence. Motions do not have to be programmed sequentially; up to 40 parallel commands can be accommodated.

The state diagram for any mechanism is shown in Fig 7. The mode of control of the functions of any manipulator mechanism varies. Simple mechanisms require a timed execution only while others use intermediate limit switch locations or counting functions. The monitor routine must be flexible to allow various combinations of control mode to be effective. In the more complex movements all usable control modes may be required.

The monitor functions at approximately $25H_z$. On initialisation the state of the limit switches is interrogated. A cross reference table is used to link the limit switch addresses to the placement and function of each switch. The state transition method is used to establish the initial state of each mechanism. When any command action is executed the monitor will update the state of the mechanism in accordance with the transition rules. In this way 40 mechanisms can be uniquely defined based on only 63 available limit switch inputs and the past history of movements that have occurred.

Other transition file commmands allow for commenting, conditional branching, operator decision intervention, sub-file calls and labelled entry point locations. During the operation of any manipulation sequence the operator may halt the motion and elect either to continue or abort by means of a portable hand control box.

2.2 Hydraulic and Pneumatic Controls:

As previously noted (section 1.2) the hydraulic and pneumatic control circuits housed on the manipulator allow a controlled applied force to act, yet have both a solid reverse lock and self adjusting feature. A typical control circuit is illustrated in Fig 8. The circuit consists of conventional double acting control valves with cross piloted check valves and individual pressure control. Where identical mechanisms must be controlled separately (e.g. right and left leg clamps), common pressure control is used. For control of the head clamp, a single acting circuit is used; the opposing side circuit unloading the relative pilot check valve. Activation of the rear leg position transfer modifies control of the rear stifle actuator to maximum constant force mode without a non-return check, thus permitting unloading of pressure on the stifle as the leg is raised.

Pneumatic circuits are used in less stringent applications and allow compact packaging and more flexible delivery lines. The control of carriage traverse and cradle rotate is by means of conventional servo hydraulic circuits.

2.3 Manipulator Performance:

The experimental sheep manipulator has successfully demonstrated the technical feasibility of automated sheep manipulation and enabled fully automated mechanical shearing to be proven. As an ongoing research tool the manipulator will be used to refine restraint and manipulation techniques. The present manipulation sequence takes aproximately ten minutes to complete, will accommodate sheep from 30 kg to 45 kg body weight and allows access to approximately 80 per cent of the wool. Ongoing research will establish upper and lower bounds to the size range and accessibility of the equipment. Simplification of the manipulation sequence will be examined as a more dextrous shearing robot becomes available.

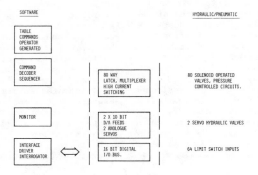

Fig 6. Block Diagram of
Manipulator Control System

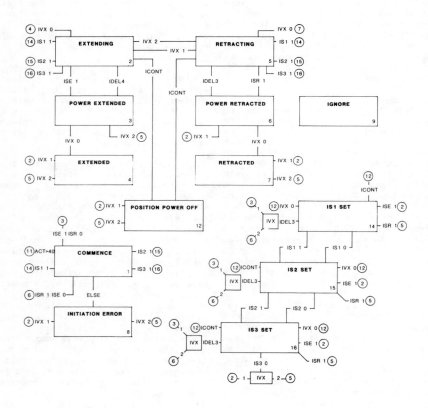

Fig 7. State Diagram for Actuator Control

3 Conclusion

Control techniques and restraint mechanisms developed for automated shearing have numerous applications to other industries.

The state variable based controller may be applied to any machine that requires flexible parallel programming of interacting events with differing modes of control such as automated packaging plant, minor assembly tasks, automated jigs and fixtures, work tables for industrial robots and conveyor systems. The benefit of this approach over hard wired logic sequencing lies in the flexibility to change or alter sequences simply by editing command files; a feature essential when developing equipment such as the manipulator in a research environment.

Sheep restraint mechanisms have obvious application for animal husbandry and control. The technique of firm yet comfortable restraint may have application to the veterinary or slaughter industries.

4 Acknowledgements

The work described in this paper was supported by the Wool Research Trust Fund at the recommendation of the Australian Wool Corporation. The authors wish to acknowledge the help received from members of the research team and from the Department of Mechanical Engineering of The University of Western Australia.

5 References

Ewbank, R. 1968. The behaviour of animals in restraint. Abnormal Behaviour in Animals. M.W. Fox ed. Saunders Co., Phil.

Key, S.J. 1983. Automated restraint and manipulation platform. Report on Driver Software Status. Technical report 221-R-6/83, University of Western Australia, Department of Mechanical Engineering.

Leslie, R.A. 1980. Preliminary report on animal manipulation. Technical report 200-R-80, University of Western Australia, Department of Mechanical Engineering.

Trevelyan, J.P., Key, S.J. and Owens, R.A. 1982. Techniques for surface representation and adaptation in automated sheep shearing. Proc. of 12th International Symposium on Industrial Robots, Paris.

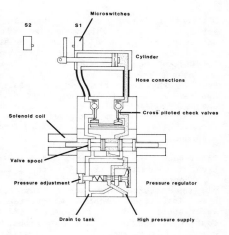

Fig 8. Typical Manipulator Hydraulic Circuit

JAPAN'S TECHNOLOGY FARM

Noboru Kawamura
Member ASAE

Distinctive features of the Japanese farm are its intensiveness and small scale; yet it is highly mechanized with small farm machines. But its product cost, for example, of rice, the staple food of the people, is three to four times higher than international rice prices due to higher labor costs, inefficient mechanization of small scale operations, and the fact that 87% of Japanese farmers are part-time. Japanese farm productivity lags behind that of the industrial sector. Large scale farming techniques cannot be applied because of limited farming area and highly industrialized socio-economic conditions. Japanese agriculture should be reformed to be competitive in the open market and to adapt to labor shortages by applying new technologies such as computerized crop growing, robotics and intelligent machines.

The tendency to use robotic and intelligent machines began with the application of automatic control to the farm machines. It was mainly simple one input-output feedback control systems, and is gradually developing to multiple inputs-outputs systems for the total control of machines using computers. Intelligent machines may be rather easily introduced for intensive and labor short farming; or dangerous and troublesome jobs; for ease, safety and comfort of operation; and fool-proofing. Labor saving is the most powerful method of cost reduction. Operation without a man or multiple machines operated by one man have many advantages. Intensive farming needs more precision operations to increase the productivity of adapting plants and environments; therefore, much information must be used to operate machines. Ease of operation and fool-proofing for the operator are necessary for big machines with greater operating width and higher speed, as well as for unskilled operators in part-time farming even with small machines. Safety and comfort of the operator are among the most important design factors for farm machines.

Another aspect which will help promote the introduction of robotic and intelligent machines is the development of the electronic and computer industries. Applications of electronics and computers to farm machines must be simple and cheap, not complicated and expensive. Electronic elements and computer prices are decreasing with the rapid development of the electronic industry and are coming within reach of agriculture. Another aspect is that farms and dealers must have some knowledge about electronic and computer technology. Training of dealers' technicians and implemention of diagnostic instruments for electronic systems must be accomplished by the dealers.

Farm machines are expected to incorporate robotic and intelligent machine elements in the future. Machines may be categorized into two categories on their development; one is the ordinary farm machines such as tractor and combine equipped with intelligent elements; the other is new concept of machine for robotization of farming such as gantry or driverless machine.

The author is: Noboru Kawamura, Professor, Agricultural Engineering Department, Kyoto University, Kyoto, 606 Japan.

Japanese farmers, in spite of small scale operations and part-time farming, have introduced many farm machines with electronic or computer control. Tractors with electronic-hydraulic control hitches, electronic controlled combines and microcomputer control dryers are now used. Most of these apply simple on-off control instead of expensive servo control. Although they are simple systems, they must handle plants, biological materials and soils, and also operate on open fields where many disturbances occur. Here exists the difficulties of development of robotics and intelligent machines for agriculture. In the following some research on the technology farm using intelligent machines and practically-used automatic control machines are discussed.

TRACTORS

Almost all tractors have hydraulic draft control of three point hitch that was the first successful application of automatic control to the tractor, but in Japan electronic-hydraulic control hitch for the rotary tiller has been used. Ordinary draft control applies one input-output system by sensing the upper link force with a spring, and changing plowing depth with hydraulic cylinder, but there are two causes of changing plow draft; that is, tilling depth and soil condition. It is impossible to identify the causes only be detecting one input; therefore, two sensing devices and judging elements of a logic circuit or computer are necessary. For the rotary tiller there are two causes to change its tilling torque, and two outputs or control operations to change tilling depth and pitch.

Figure 1 shows our research on microcomputer control system of a rotary tiller on a hydro-static drive tractor (Fujiura et al. 1982). The rotary tiller was suspended by a torsion-bar and its deflection due to the tilling reaction forces was detected as tilling torque with a linear variable differential transformer. Tilling depth was detected by a gauge wheel (Fujiura et al. 1980). A block diagram of the control system is shown in Fig. 2. The following signals were detected and each was input to the microcomputer:

(a) Swash plate angle of HST: it was nearly proportional to tractor speed
(b) Tilling reaction force: torsion-bar and LVDT
(c) Tilling depth: gauge wheel and LVDT
(d) Height of three point hitch: LVDT
(e) Speed of engine: magnetic speed sensor.

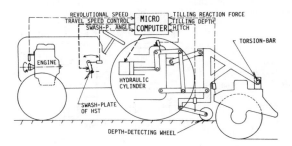

Fig. 1   Tractor with Microcomputer Control

The microcomputer was an 8-bit parallel processor (MB8861) which was equivalent to the MC6800. The control program was made of machine code and its size was about 1.4 k Bytes. Each analog signal was selected sequentially by multiplexer and was converted to digital signal by an 8-bit A/D converter. Engine speed was measured by the number of machine clock pulses of microprocessor unit (1 MHz) during one revolution of crankshaft which was detected by an electromagnetic sensor. By the interrupt program, which was done every one revolution of engine, engine loaded torque T was calculated using memorized torque-speed characteristic curve in the processor and following equation:

$$T = T_e - J \cdot \dot{\omega} \tag{1}$$

where J is inertia moment of the rotating part of engine, $\dot{\omega}$ is the changing rate of engine speed, and T_e is engine torque memorized in the processor. Output signals actuated solenoid valves of cylinders of the three point hitch and of the HST.

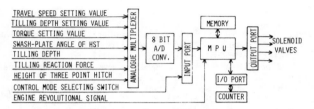

Fig. 2 Block Diagram of Control System

Five program control modes were provided for choice and tested on the paddy field. Mode 1 was to control the tilling depth and pitch by giving their setting values. Mode 2 was to control the tilling depth and pitch by judging the causes of tilling torque fluctuation. In case torque was changed by pitching or sinkage of tractor, tilling depth was controlled; in case torque was changed by change in soil hardness, tilling pitch and depth were varied to keep torque constant. Mode 3 was to control the tilling pitch in response to the tilling reaction force, and the tilling depth in response to detected depth by gauge wheel. Mode 4 was the same as mode 3, but the travelling speed was controlled by detected engine speed. Mode 5 was almost the same as mode 2, but the engine torque was computed with eq. (1) by detecting engine speed. Each control mode had advantages for flat tillage on paddy field and deeper tillage on upland, and also for more precise land preparation.

Fig. 3 shows an elecronic-hydraulic control for a rotary tiller that had come on the market. Upon sensing the engine speed by an alternator and tilling depth by a potentiometer attached on a lifting arm shaft, the depth of rotary tiller was controlled with these two inputs.

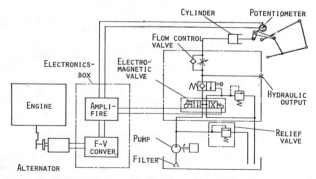

Fig. 3 Electronic Hydraulic Control System of Tractor

Tractors with electronic position control or level control of rotary tillers have been used in Japan. The position control was done by detecting its depth at the rotary tiller cover. Level control was done by sensing the inclination of a pendulum with a photo-electronic unit mounted on the rotary tiller or the tractor. The length of lift rod was changed with a hydraulic cylinder. This system is useful for the land preparation of paddy field or on hill side.

There are many possibilities for applying computers to the tractor to control engine performance, active seat suspension, automatic steering etc. Some research work on the automatic steering of tractor has been done, but it has not yet come to practical use.

PEST CONTROL MACHINES

Japan's Industrial Robot Association envisions widespread robotization of farms for chemical spraying within this decade, because it is a dangerous operation for the operator by either touching and breathing the chemical. For safety, driverless operations of a field sprayer and air blast orchard sprayer were investigated using microcomputer control. Figure 4 shows an experimental self-propelled orchard sprayer which has two photosensors A and B and microcomputer (6800). When the photosensor A detected a tree trunk, spraying started; and when sensor B detected it, spraying stopped. For the purpose to develop remote operation by host computer, the location of the sprayer in the orchard which was detected by sensor A and the travelled distance calculated in computer was memorized, and also the deposits of chemicals on leaves were transmitted to the microcomputer with FM signals (URA et al. 1981).

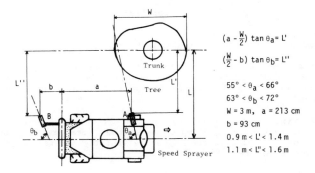

$$(a - \frac{W}{2}) \tan \theta_a = L'$$

$$(\frac{W}{2} - b) \tan \theta_b = L''$$

$$55° < \theta_a < 66°$$

$$63° < \theta_b < 72°$$

$$W = 3\,m, \quad a = 213\,cm$$

$$b = 93\,cm$$

$$0.9\,m < L' < 1.4\,m$$

$$1.1\,m < L'' < 1.6\,m$$

Fig. 4 Self-Propelled Orchard Sprayer

Pest control sprinklers for the mandarin orange on terraced orchard have been widely used in Japan for safety and labor saving, because the remote control of spraying was easily done from the pump station and droplets from sprinkler were relatively large, resulting in the decrease of drift of chemical. But this system was very expensive even though it was used for irrigation (Ura et al. 1980). Figure 5 shows our research self-propelled pest control sprinkler in which travelling and spraying was controlled by microcomputer (Kawamura et al. 1982). It was suitable for small orchards and was a cheaper system than the pipe line system. It was driven by a 1.0 kW air cooled engine (4), and travelled along a guide PVC pipe (20) laid on the ground. Chemical liquid was supplied from the main pipeline to the sprinkler head (8) with a hose. It started with an automatic clutch and stopped travelling to spray by detecting a tree with an infrared photosensor (13) or by counting the distance of travel with sprocket revolutions. The microcomputer was the same as above (6800). Interrupt was caused by each pulse from the revolution sensor (15), and the number of pulses was compared in IRQ interrupt program to the standard count numbers (150 pulses, about 7 m). When the photosensor detected a tree, counting of pulses was stopped, and the sprinkler stopped and sprayed chemical with rotating of sprinkler head several times. If there were no trees ahead for 7 m or 150 pulses, pulses were counted to the maximum count number (210 pulses, about 9 m), and there it stopped travelling and sprayed chemical. The sprinkler head sprayed chemical only to the upper and lower terrace sides where trees were planted. Chemical consumption and drift were reduced for safety through the use of such a spraying system which detected the target tree and sprayed to it. At the end of a lateral branch it stopped and went back to the main pipe line winding a hose on the drum, and then moved to the next lateral.

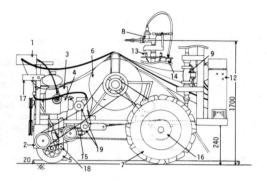

Fig. 5 Self-Propelled Sprinkler

COMBINES

Many combines equipped with automatic steering, cutting height and threshing control by electronic systems have been used in Japan. At the beginning of our research it was thought that a few percent of the combines might have automatic control systems for optional use, but in reality 90% of relatively large size combines--they might be the smallest ones compared to USA ones-- are equipped with electronic control system. The reasons why so many electronic control systems were accepted by farmers were ease of operation and lower grain losses, and also the higher reliability of electronic components supported by good maintenance by the dealers.

Figure 6 shows a driverless, full automatic control combine with a microcomputer. Sensors A, B and C in front of the cutter-bar were, respectively, a control switch, steering switch and auxiliary switch. When the combine came into the field to harvest, sensor A at first contacted plants, initiating operation of electronic control systems. Also, the cutting mechanisms began to operate. Sensor B operated thenceforth to steer automatically along the row of plants. When the combine came to the end of the field, sensors A and B were switched off and a sequential control was begun to turn the corner. This control, computer timed, programmed the combine to travel some distance straight, then the tail of combine shifted outward from plant rows, then the combine turned left, then moved rearward while continuing to left, coming at approximately a right angle to the plant rows. Then the combine moved forward to harvest in the transverse direction. Repeating such automatic steering and sequential turning controls, it continued to harvest all the field, and then stopped. If the steering was not correct and some plants were left unharvested, sensor C touched their stalks and initiated another sequential control to correct travelling direction, turn slightly to the right and move forward steering to the right so as to locate the unharvested plants in front of the cutter-bar. The combine was equipped with cutting height control and threshing depth control. All of the controls were performed in on-off mode. Even though only three inputs may be used, there are many possibilities of doing field operations with intelligent machines.

Some combines were investigated using photosensors to assist the above-mentioned control. Photosensors were used for identifying the distance between plants and the combine during turning. If irregular slippages occurred during turning, the combine failed to turn using only sequential control. With photosensors, the combine had more reliability and flexibility to adapt to soil and field conditions.

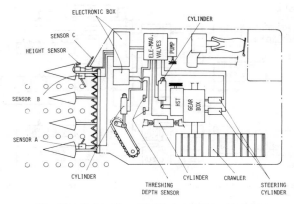

Fig. 6 Driverless Full Automatic Combine

The performance of the combine differed with plant conditions of variety, maturity and moisture content, so that adjustment of the combine was difficult. An adaptive control system which controlled travelling speed and the setting values of feed rate by logical operation of the three inputs of engine load, feed rate and its fluctuation is shown in Fig. 7 (Kawamura et al. 1975, Kawamura et al. 1978). Combines should be operated at the optimum feed rate of the threshing cylinder and without its speed decreasing. It might be possible by mounting a bigger or additional engine to overcome the overload of the cylinder, but a small combine, especially for soft paddy field, should be constructed lighter by mounting a lighter engine.

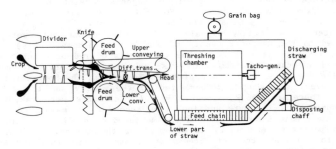

Fig. 7 Adaptive Feed Rate Control of Combine

The control systems were tested with wired logic system or microcomputer. Three input signals were cylinder speed, feed rate detected at the conveying device and its memorized signal for 3.5 seconds which was the equivalent interval for material to be conveyed from the detected point to the

57

cylinder. The feed rate fluctuation was judged by comparing its presently measured signal and a memorized value. A logic circuit or the computer used the three signals and judged whether the feed rate was adequate, too much, or too little, referring to memorized engine torque-speed characteristic performance. The speed of the combine was controlled with HST. Values of feed rate were set by a potentiometer and DC motor as shown in Fig. 8. In the case of the wired logic circuit, simple digital signals were compared with setting values which were supplied to the logic circuit. A combine with this control system was able to operate without any consideration of plant condition. It selected optimum setting values of feed rate automatically, harvested faster, and gave less grain losses than a manually operated combine.

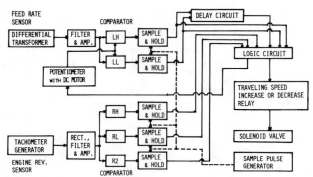

Fig. 8 Electronic Circuit of Adaptive Control of Combine

There are also some investigations of the use of synthesized voice for warnings and for queries of what to do for safe operation and easy handling.

FRUIT HARVESTING ROBOT

A self-propelled fruit harvesting robot is now being investigated in our laboratory (Fig. 9). It consists of battery car, articulated manipulator with five degrees of freedom as shown in Fig. 10, television camera and an 8 bit microcomputer. Figure 11 shows the block diagram of image input to the computer. The television camera has MOS type image sensor and identifies color differences of fruit and leaves. The MOS image sensor has 384(H) x 485(V) picture elements and the output signals consist of Y_e (yellow), G (green), W (white) and C_y (cyan). For this experiment two signals of red color $R = W + Y_e - C_y - G$, and brightness $Y = W + G = Y_e + C_y$ were input to the computer. Two binary signals of R and Y as one byte were directly transferred to the computer memory by DMA (direct memory access). Transfer speed was 4.5 μs. It took more time if signals of red or other colors were transferred to the CPU from the pulse of microcomputer. To reduce the memory size and time, every four picture elements in the horizontal scanning line and every six scannings were fed to the memory and the others were omitted. Therefore, 96(H) x 81(V) = 7,776 elements were taken, and the pattern memory was 1 k Byte and took 1/60 second to be memorized. The distance of the fruit from the camera was measured comparing the difference of memorized patterns between 15 cm movement of the camera. Treatments were included to transform the reflected shine spots on the fruit surface to the surrounding color.

The manipulator was driven from the signals of the computer by six DC motors shown in Fig. 10, and its five angular positions were detected with potentiometers. Fingers of this manipulator were designed to grasp a fruit. If the fingers failed to grasp a fruit, it was detected by a limit switch at

58

the fingers' base, and grasp motion was repeated. The manipulator was made to grasp a fruit, but for practical use we are planning to use a cutting device and a catching pocket with a flexible chute.

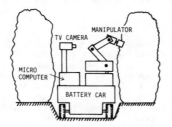

Fig. 9   Fruit Harvesting Robot

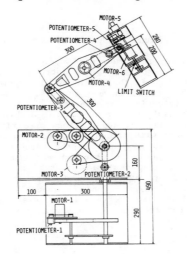

Fig. 10   Manipulator

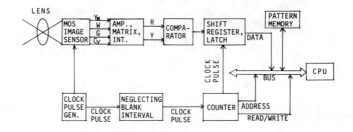

Fig. 11   Block Diagram of Image Sensing Device

Experiments were done to harvest tomatoes. It was found that most of the functions were fairly adequate, but the operating speed was slow because it used an 8 bit microcomputer and small DC motors. It must be cheap and simple for agricultural use. Other problems to be solved in the development of the fruit harvesting robot were, for example, to identify the fruit hidden behind leaves or other fruit and differences of color between mature and immature fruit. Much information about fruit color, shape and hardness should be obtained.

Another robot experiment was done using a simple image sensor with a 64 x 64 matrix and red, blue and yellow color filters. The camera consisted of two lenses; the later one was set to be in focus with the focal plane of the front one to be able to approach close to the object fruit. The camera was attached near the fingers in this system, and the manipulator moved to the position of the fruit which was memorized at a travelling position of the manipulator. The distance and size of fruit could be calculated from the image size increment by comparing two points of the manipulator's approach to the fruit. The accuracy of manipulator movement was better than the former system, because the size of the image became larger upon approaching the fruit.

COMPUTER OPERATED GANTRY

A computer operated gantry that runs over a plot of land and does everything from tillage to final harvesting has been introduced and tested at the Agricultural Research Center of the Ministry of Agriculture, Forestry and Fishery in Tsukuba. Figure 12 shows the overall view of the gantry and in Table 1 the main specifications are shown. The gantry was attached with rotary tiller, rice transplanter, pest control machine and combine for each operation. Their operations and applications of irrigation water and liquefied fertilizer were done with the program of a central computer. The gantry was also programmed to go out from the warehouse to each plot and to return. All operations could be observed with television in the central station. Signals between the central computer and gantry were transmitted with optical fiber cable to avoid noise signals and to transmit many signals. It transmitted 131 signals and 24 channels. This gantry has been built for research and is directed toward the development of an agricultural robot for farming of the future. A small gantry for field experiments was also built.

Fig. 12   Computer Operated Gantry

Among the advantages were that no man entered the field and therefore no soil compation occurred, and day and night operation was possible, resulting in higher field capacity. A rice growing test last year resulted in a yield 19% above the national average, and more uniform growing of rice and better soil conditions were observed. The cost comparison of gantry and farm machine system is shown in Fig. 13. The cost of the gantry system was about 1.6 times as expensive as an ordinary farm machine system. If the concrete rail at the levee was changed to a free running road and the gantry was built of lighter construction, there would be a possibility of introducing the gantry system to intensive farming in the developed countries.

Table 1. Specifications of Computer Operated Gantry

Item	Specification
Field Plot	5 a x 4
Gantry Crane Spane	12 m
Gantry Crane Height	3.6 m
Gantry Crane Weight	9,000 kg
Speed	0.1-0.4 m/s
Control	Central Computer Control
Signal Transmission	Optical Fiber Cable
Visual System	ITV Camera and Monitor TV
Artificial Rain	2,500 1/h

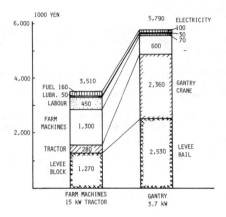

Fig. 13 Cost Comparison Between Farm Machine and Gantry

The greenhouse is a suitable place for the gantry system to be introduced. A gantry like this one had already been introduced in the greenhouse at Shimane University. It spanned the width of the greenhouse and was used for transportation, pest control and irrigation. It was operated by computer with a program that was varied with the stage of plant growth. All operations were observed by ITV camera. Controlled greenhouse functions were temperature, light, humidity and CO_2 gas to give optimum conditions to the plants in the closed environment. The computer has wide adaptability for plant growing in such a situation because of its capability for handling many conditions of soil moisture content, plant growing and temperature etc. Many of the control elements with microcomputers for greenhouse cultivation had been used in Japan. Those tendencies would expand to the open field intensive farming in the future with simple coverage of plastic film and plastic film mulching, etc.

For the future work of the technology farm, robotic and intelligent machines should be investigated to construct systems which are simple and cheap but with higher reliability. In addition, much information on plants and environments should be collected and their correlations would be analyzed for effective use by the software of intelligent machines.

REFERENCES

1. Fujiura, T., N. Kawamura, and P. SiGia. 1980. Automatic control of rotary tilling tractor (Part 2). Research Report on Agricultural Machinery, Kyoto Univ. 10:1-14.

2. Fujiura, T., and N. Kawamura. 1982. Automatic control of rotary tilling tractor (Part 4). Research Report on Agricultural Machinery, Kyoto Univ. 12:1-13.

3. Kawamura, N., K. Namikawa, M. Yukueda, T. Fujiura, and T. Kawamura. 1975. Automatic feed rate control of combine in two inputs system and its adaptive control. Memoirs of the College of Agriculture, Kyoto Univ. 107:1-35.

4. Kawamura, T., N. Kawamura, and K. Namikawa. 1978. Adaptive feed rate control of head feeding type combine. Research Report on Agricultural Machinery, Kyoto Univ. 8:68-96.

5. Kawamura, N., M. Ura, and N. Kondo. 1982. Microcomputer control of self-propelled pest control sprinkler. Research Report on Agricultural Machinery, Kyoto Univ. 12:31-37.

6. Ura, M., and N. Kawamura. 1980. Fundamental study on pest control systems by high pressure type sprinkler in the orchard. Research Report on Agricultural Machinery, Kyoto Univ. 10:26-45.

7. Ura, M., and N. Kawamura. 1981. Microcomputer control of field and orchard sprayers. Research Report on Agricultural Machinery, Kyoto Univ. 11:27-46.

APPLICATION OF AGRICULTURAL ROBOTS IN JAPAN

Nobutaka Ito

REVIEW AND PRESENT STATUS OF MECHANIZATION

During the past three decades since the end of the World War II, agricultural mechanization in Japan has progressed with rapid industrialization.

In the first decade the replacement of the manual work with mechanical operations removed the animal from farming operations.

The tractor came first from the introduction of small scale tractors such as power tillers or two wheel tractors equipped with rotary tillers, which were mostly walk-behind machines. This led to the development of the four wheel tractor of 15 to 20 horsepower with two wheel drive. Now, most tractors are four wheel drive. Also, some of the crawler type tractors are equipped with three point hitch systems and PTO drive shafts. In parallel with this tractorization, the rice combine was successfully developed and mechanization in rice harvest began in the second decade. A machine equipped with only cutting and binding functions was distributed for a short time, and is still used in some mountainous areas of Japan. It is well known that Japan's stable crop is rice, hence mechanization was started from rice production. Also, prior to the successful development of rice combine, the transplanter was made and is widely used today.

Rice mechanization is almost totally complete from transplanting to harvest and the post harvest process. The direct seeding method of coated rice is now gaining acceptance. This technique will be widely accepted by the farmer for the following reasons; (1) a high percentage of germination, (2) better weed control by use of effective herbicides and (3) lodging resistance because of the undergrond seeding. Automation is one of the themes in addition to safety and energy saving in developing Japanese agricultural machinery. Some of the current commercial agricultural machines are being equipped with microprocessor-based control systems; however, optimal design and functions should be for promoting automation to get better efficiency and reducing the waste operation and handling. Some rice combines are equipped with turn table mechanisms for the purpose of reducing traveling distances and saving time for turning in harvesting operations. Some combines are equipped with hulling functions in addition to the conventional combine functions.

VARIOUS CONTROL APPLICATIONS TO TRACTOR

The basic philosophy for automation of tractors is based upon the following:

(1) to improve control in operation
(2) to increase efficiency and performance in operation
(3) to increase accuracy in operation
(4) to save energy

---

*The author is: Nobutaka Ito, Associate Professor, Agricultural Machinery Department, Mie University, Japan

(5) to make complex and combined operations easier
(6) to increase safety

For the basic functions in tractor-implement control, the following controls will be desired to meet the requirements listed above:

(1) position control
(2) posture control
(3) draft control
(4) forward speed control
(5) steering control (automatic guidance control)
(6) monitoring and warning functions

As described before, four wheel drive tractors are more popular than two wheel drives, because of the higher efficiency of tractive force, and better trafficability and maneuverability in paddy fields. Most of them are generally less than 30 horsepower.

In the past, there were two typical controls for implement-position and draft control. Both are still in use. They consist of mechanical linkage with sensing devices installed in the tractor. A top link is used for sensing the compression force. Recent control systems consist of electronics and hydraulics which are quite different from the previous mechanical system.

Based upon the basic philosophy and required functions for promoting automation, some examples of controls for tractor-implement systems are introduced.

Figure 1 shows a current control system for tractor-implements, which is already commercialized (Sakai 1982). For the signal to control the implement position, the rotative angle of three point linkage lift arm is sensed with the rotative type potentiometer. Depth control is done by keeping the angle of soil cover constant as shown in Fig. 1. Draft force is sensed by mechanical method, but for precise control, the displacement or deflection of the sensing spring attached to the upper link is converted to an electrical signal. In addition to this, a limit switch for sensing the upper-link limit of lifing is installed.

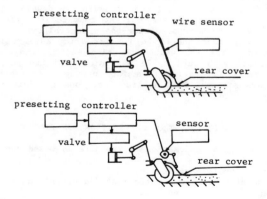

Fig. 1 Mechanical (up) and Electrical (down) Depth Control

Those signals produced from each sensor are supplied to a controller which makes a decision whether to lift or lower the implement. A solenoid valve then opens the circuit to actuate a hydraulic cylinder in the appropriate direction. The lowering speed of the implement can be adjusted by the flow control and check valve installed between the actuator and the solenoid valve as shown in Fig. 2. This is an example of position, draft and depth control by use of one controller (Akama 1982).

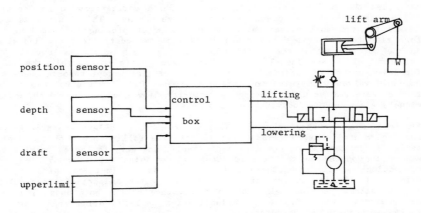

Fig. 2   IC Control Block Diagram for Position and Depth Control

Figure 3 shows a horizontal level control system for the implement mounted on tractor. The purpose of this system is to maintain the implement at level or any any given angle irrespective of the tractor posture during an operation such as rotary tillage, land leveling, puddling, etc. Rice land preparation needs this fine leveling.

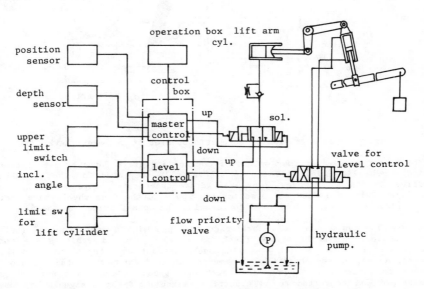

Fig. 3   Schematic View of Level Control for Implement

The principal difference from a conventional three point linkage is that one of tractor's hitch lifting rods is replaced with a hydraulic cylinder with a stroke sensor.

The rolling angle of the tractor main chassis can be detected by the sensor and then the lift rod containing the hydraulic cylinder extends or contracts depending on the electrical signal in proportion to the rolling angle. This maintains the implement at the horizontal which corresponds to the control for maintaining the implement parallel to the ground. This control is necessary for the soil pulverization operation after plowing and puddling before rice transplanting. As shown in Fig. 3, two controllers are installed in one box to control the position and tilling depth of the implement. Two solenoid valves are installed; one for master cylinder control, the other for level control. Position control can be precisely done through these two solenoid valves with pulse modulation control. Oil flow control can be led to the required actuator through a priority flow divider depending on the signals given by the individual controllers.

Figure 4 shows another schematic diagram of horizontal level control system for implements) (Kisaka (1982). It detects the rolling angle of the implement. It is mounted on the implement itself, however, for newer models, the sensor is mounted on the tractor chassis. By doing this, inconvenience in connecting and disconnecting the signal wire during implement hitching was eliminated. Inside the sensor a contactless switch and anti-freeze coolant are installed. A float with the metal attached at the both ends floats on the coolant. About 0.5 deg. of inclination angle (rooling angle) of the implement-tractor can be detected.

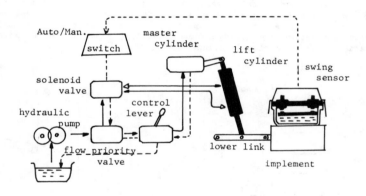

Fig. 4 Schematic View of Automatic Control of Implement Leveling

The author studied another method of controlling draft force by use of sensing wheel slip (1975). This was successfully completed for the two wheel drive tractor. However, four wheel drive tractors are more popularly used today. This idea does not practically apply to the four wheel drive tractor since no wheels can be used for sensing the actual traveling speed.

Figure 5 shows a commercial bulldozer equipped with a draft control (bulldozing force control) system (Yoshidda 1982). Tracked vehicles are more difficult to sense actual traveling speed without an extra wheel. In this case, a doppler sensor (10GHz) is mounted at the rear part of the operator's seat to estimate distance. The rotational speed of the drive sprocket is obtained and converted to an electrical signal. The two signals are used for calculating slippage.

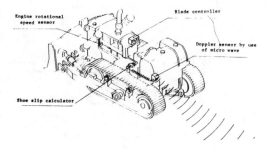

Fig. 5 Bulldozer Blade Control

If the computed slippage is less than the preset slippage, it means the load
acting on the bulldozing blade is less than the allowable load; however, if
it is larger than the preset slippage, the bulldozing blade should be lifted
up to reduce the load; then the slippage can be decreased down to the preset
slippage.   Therefore, the slippage can be varied depending on the change of
the load acting on the bulldozing blade. Knowing the variation of the slippage and specifying a range, the bulldozing blade force can be controlled.
Figure 6 shows the schematic block diagram of bulldozing blade control.

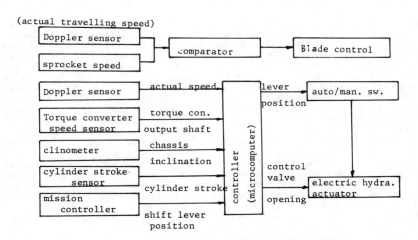

Fig. 6 Bulldozer Control System with Microcomputer Mounted

In bulldozing operations, two controls are needed; leveling control and load
control for controlling the load acting on the blade.   For the leveling
control a laser beam is used.   It is considerably more expensive, but the
leveling accuracy is ± 2 cm at 10 km/hr traveling speed or ± 5 cm for 3 to 4
hundred meters traveling distance.   The ground should be almost flat for the
bulldozer to overcome and negotiate the soil obstacle. A load control function is needed to complete the work smoothly.   There seem to be few bulldozers equipped with both load and leveling control systems. Figure 7 shows
the schematic block diagram of the combined system of level and load.   This
combined control system has the following advantages:

(1) simple structure and no special sensor for detecting the load is needed
(2) low cost
(3) presetting of the load and position of the blade can be easily done by adjusting the unloading valve and potentiometer

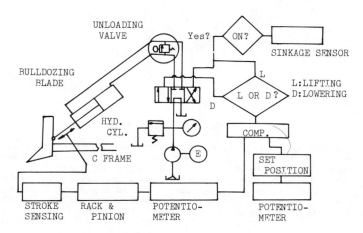

Fig. 7 Bulldozer Blade Control for Leveling and Bulldozing Load

Prior to the bulldozing operation the load and lift cylinder stroke are preset. By doing this the cutting depth is determined as the bulldozer travels, the soil volume being pushed by the blade and the soil cutting resistance gradually increases. Simultaneously, the hydraulic pressure in the lift cylinder (piston side) becomes higher. As that pressure becomes higher than the preset pressure value of unloading valve, oil is released through the unloading valve and flows into the rod side of lift cylinder. At this moment the cylinder is displaced, lifting the blade. The lift cylinder sensor detects the displacement of this cylinder immediately and a solenoid valve opens so the oil from the hydraulic pump can be supplied to the piston side of the cylinder. If the piston side's pressure becomes equal to the load pressure, the unloading valve can be closed, then the lift cylinder pushes the blade to the present position. As soon as it returns to the original position, the solenoid valve also returns to the neutral position. To protect the bulldozer from overload, a sensor for detecting the clearance is mounted. In case this sensor is activated, the controller lifts up the blade and overrides any other function concerned with the bull-dozing operation. Bulldozers are being accepted by some farmers because of high tractive performance and powerful operation in addition to more sta-bility. Bulldozers can be a great help in reclaiming land and rejuvenating soil for cultivation again. It can be used as a crawler tractor in ordinary farming too. From this point of view, the bulldozer equipped with PTO and three point hitch system is increasing.

Microcomputer-based control systems can be seen in other construction machines. Figure 8 shows the schematic view of a large scale power shovel control system. (Tanaka et al. 1981). It consists of electrical operation lever and pedal sensors installed in the chassis, as well as electric hy-draulic servo valves for controlling the hydraulic components, solenoid valves, and a microcomputer. In correspondence with the operator's handling of the operating lever, the microcomputer controls the following operations:

(1) Oil flow control corresponding to the operating lever angle.
(2) Combination control of hydraulic pumps and valves; depending on operation, the oil quantity from multi pumps can be controlled in the order of priority and only the required volume of oil can be delivered.
(3) Bucket posture angle can be maintained at any given posture until the dumping operation is completed. This can be done through the microcomputer by use of three angle sensors attached to the boom, arm and bucket.

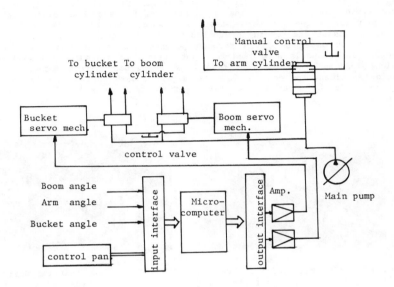

Fig. 8 Schematic Diagram for Automatic Digging and Loading for Front Loading Type Power Shovel

RADIO CONTROL

The author applied radio control to rice combines (1977). It was successfully operated in the paddy fields. In that study radio control was used to assist the turning of the vehicle at the end of the rice field. While the combine was being operated along the row with certain cutting width; steering, cutting height and feed rate controls guided the machine automatically. During this operation, the operator only monitored. The same thing can be seen in the current manufacturing plants. Only one person supervises five to ten controlled tools. If they are working without any trouble the supervisor does not have to do anything but watch the warning panel to tell him which tool is in trouble. The merit in radio control is that it releases the farmer from the hard physical work in harvesting. However, the total time required for completing the harvesting operation is the same as for a manual operation. For this problem, multi radio control for two or three different machines can be considered. But, it looks impossible for the Japanese farmer to adapt this based upon his small scale farm. With construction machinery the application of the radio control could be useful and effective, because there is danger associated with the construction environment.

PROGRAMMED CONTROL

The author initiated the application of programmed control to agricultural machinery in 1974. It was applied to the steering control of rice combines and also applied to the loading unit for engine endurance testing (Ito 1975). The author's original purpose was to complete the automatic guidance control for rice combine, combining the three controls previously described and this programmed control.

Figure 9 shows the whole view of the combine equipped with programmed control system for steering. The application of the same control to the loading unit for engine durability test is shown in Fig. 10. This was the first trial to assist the turning of combines when they come to the end of a rice row. Input data was given through punched tape which was manually produced. Two channels were used for controlling two hydraulic solenoid valves which produced right and left turn in steering and uniform motion. In the field experiment the combine followed expected path satisfactorily which was dependent on the punched input tape. It was not operated in an actual rice field.

Fig. 9 Combine Equipped with Programmed Control for Steering

Fig. 10 Loading Unit for Engine
Under Programmed Control

DRIVERLESS COMBINE

In 1981 the author and his graduate student constructed the microcomputerized driverless combine (Koto et al. 1981). A driverless combine was successfully developed by one of the farm machinery manufacturers in Japan, Iseki, in 1976. Since that time the importance of electronics for agricultural machinery automation was closed up. To date the driverless combine has not been commercialized because of the following reasons:

(1) difficulty in trouble shooting for the farmer or dealers
(2) quite expensive
(3) problem in maintenance and serviceability

Some of the individual controls developed in the prototype machine have been successfully adopted to commercial machines. The combine with microcomputer is commercially produced by some of the manufacturers, but is not driverless yet. Figure 11 shows the combine constructed at Mie University. The block diagram of the control system is shown in Fig. 12. The automatic steering control is shown in Fig. 13. The signals from steering sensor and the edge sensor are supplied to the microcomputer. The hydraulic cylinders attached for indivudual steering clutches engage or disengage the steering clutch depending on the output signal of the microcomputer.

Fig. 11   Microcomputerized Driverless Combine Constructed at Mie University

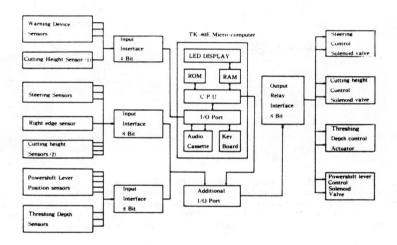

Fig. 12   Block Diagram of the Control System

The sensor for steering control senses the four conditions of "no rice plant," "right turn," "left turn" and "keep going," by use of three micro switches.

Figure 14 shows the cutting height control system. The cutting height sensor used here is similar to the one used for steering control. Because of difficulty in modifying the directional valve for lifting and lowering the cutter bar, a mini pneumatic cylinder was used for operating the cutter bar control lever. Depending on the length of the rice plant, header height should be adjusted so as to obtain the uniform supplement of rice plant with less loss. The sensors (limit switches) are attached to the entrance of the threshing unit. Header height can be controlled by adjusting the vertical feed chain conveyer driven by d.c. motors through reduction gears. For completing the full automatic guidance, the gear shift operation should be controlled. In this system a powershift transmission was used; therefore stepless shifting could be obtained. Five limit switches were installed. The location of the shift lever for changing the traveling speed was sensed with those limit switches as shown in Fig. 15. The turning motion at the corner or at the end of the rice plant row was donen under programmed control without feed back function, therefore to bring the combine back to the

71

desired location when it goes the wrong way, two sensors were used; one of them was used for controlling the travel along the rice plant row and the other was used as the swath edge sensor to detect the outside edge of the cutting width. Figure 16 shows the header control.

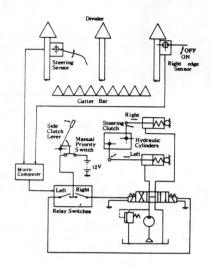

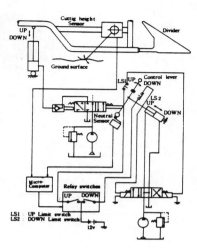

Fig. 13   Electric-Hydraulic Circuit
          Diagram Used for Automatic
          Steering Control

Fig. 14.   Electric-Hydraulic Circuit
           Diagram Used for the
           Cutting Height Control

The turning pattern to modify the programmed turning is shown in Fig. 17, whereas the programmed turning pattern is shown in Fig. 18. In the actual field of paddy, the combine equipped with this control system was tested through the harvesting operation. Automatic steering and programmed turning were satisfactorily worked in addition to the other control components.

BASIC SHAPE AND FUNCTION OF THE COMBINE UNDER JAPANESE CONDITION

To promote the application of automatic control with intelligent functions to the combine, the shape and functions should be essentially considered from the standpoint of improving the performance in saving time and shortening the distance required for completing the harvesting operation per unit area of the paddy field (Ito 1981).

For improving the efficiency and the performance in operation in the field, there may exist three tasks.

(1) Equip the special mechanism to do so.
(2) Adopt the proper method of harvesting pattern.
(3) Determine the optimum shape of the paddy field.

From the results of the combine equipped with turntable mechanism based upon the comparison with the commercial combine when it is operated in the most typical conventional way, it was found that the turntable type combine is more efficient than the conventional one in the view point of time and total distance required for the combine to complete the harvesting operation. The time and distance saved for turntable type combines can be 10% and 25% respectively for the square shape field, compared to the conventional combine.

Powershift lever
LOW — High
Hydraulic cylinder
LOW — High
R N 1 2 3
Micro-Computer
Relay switches
High Low
12V
R: Reverse N: Neutral
1: lst speed 2: 2nd speed 3: 3rd speed

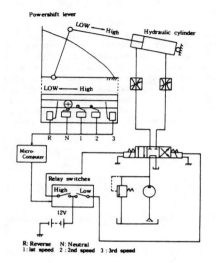

Fig. 15 Electric-Hydraulic Circuit Diagram Used for Powershift Control

Threshing depth sensors
Threshing depth control actuator (wiper motr)
Threshing drum cylinder
Feed chain
M
Trans-mission
deep — shallow
Vertical feed chain
Indicator
LS1
LS2
deep shallow
Micro-Computer
deep
Shallow
OFF
ON
12V
LS1,LS2:Limit Swit Switches
Relay switches

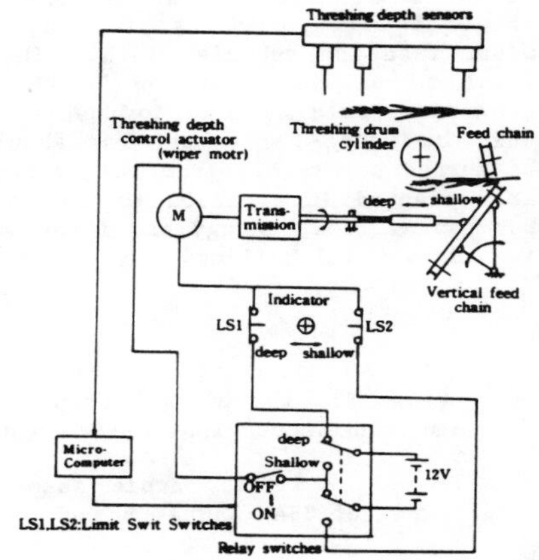

Fig. 16 Electric-Mechanical Circuit Diagram Used for Threshing Depth Control

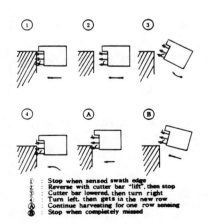

1 : Stop when sensed swath edge
2 : Reverse with cutter bar "lift", then stop
3 : Cutter bar lowered, then turn right
4 : Turn left, then gets in the new row
A : Continue harvesting for one row sensing
B : Stop when completely missed

Fig. 17 Turning Pattern of the Combine at the Corner

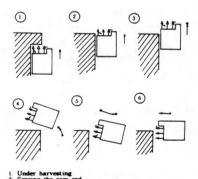

1 Under harvesting
2 Sensing the row end
3 Move forward for while, then stop and lift the cutter bar
4 Turn left, then stop
5 Reverse with right turn and stop
6 Cutter bar lowered, then gets in new row

Fig. 18 Turning Pattern of the Combine at the Corner

To increase the cutting width is more effective in improving the performance than to increase the traveling speed in harvesting operation. Adopting the turntable mechanism can simplify programming, and disturbance of the soil and ground surface of paddy can be avoided.

It can be concluded from the above discussion that the shape of the combine should be determined as follows. The width of the cutter bar should be wider than the vehicle width. The width of the processing unit should be equal to or less than the width between tracks. For crawlers, the wide grousser should be used to ensure better trafficability, and the processing unit including the cutter bar should be mounted on the turntable which can be turned separately from the tracks. The hulling function should be desirably adapted in addition to the existing functions, threshing and cutting, for saving time, energy and labor spent for the serial operations of threshing, drying and hulling.

REFERENCES

1. Akama, S. 1982. Mechatronic accessory for tractor engine. JSAM Power and Machinery Symposium Proceedings. 38-56 p.

2. Ito, N. 1981. Basic shape and function of agricultural vehicles. Trans. of JSAM Kansai Branch. Vo. 49. 33-38 p.

3. Ito, N. 1982. Basic shape and function of agricultural vehicles. Trans. of JSAM Kansai Branch. Vol. 51. 1-5 p.

4. Ito, N. 1975. Slip draft control of tractor (I). Trans. of JSAM. Vol. 37. No. 1 (No. 132). 6-12 p.

5. Ito, N. 1975. Slip draft control of farm tractor (II). Trans. of JSAM. Vol. 37. No. 2 (No. 133). 156-163 p.

6. Ito, N. 1982. Bulldozer blade control. Vehicle automation symposium proceedings. 17-18 p.

7. Ito, N. 1975. Programmed control simulator for engine testing. The Bulletin of the Faculty of Agriculture, Mie University. No. 53. 213-226.

8. Kisaka, H. 1982. Automation for tractor implement. JSAM Power and Machinery Symposium Proceedings. 72-98 p.

9. Sakai, T. 1982. Current tractor technology. JSAM Power and Machinery Symposium Proceedings. 2-18 p.

10. Yoshida, H. 1982. Hydraulic power unit. Mechanization of Construction. Vol. 14. 54-57 p.

11. Tanaka, S. et al. 1982. Mitsbishi power shovel MS580. Mechanization of construction. Vol. 11. 39-40 p.

For the reference, the electric circuit diagram of I/O interface is shown in Fig. 19.

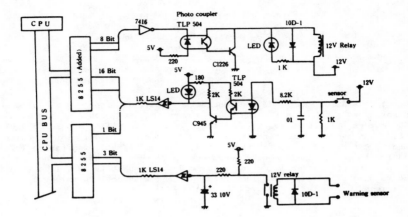

Fig. 19 Electric Circuit Diagram of I/O Interface

AGRICULTURAL ROBOTICS IN JAPAN: A CHALLENGE

FOR U.S. AGRICULTURAL ENGINEERS

W.F. McClure, Professor, PE
Member ASAE
North Carolina State University
Raleigh, North Carolina 27650

Japan, consisting mainly of the four islands Hokaido (in the north), Honshu, Shikoku, and Kyushu (in the south), has more than 500 mountains higher than 2000 meters. It is estimated that their population will reach 121,870,000 by 1985. The average farm size in 1980 was 2.9 acres, and only 16% of their land is capable of producing rice (35%), livestock (26%), vegetables (17%), and fruits (8%). Their agricultural system is very intensive. The challenge to the Japanese farmer is not only to produce in quantity as well as quality but to make full use of all they produce. And they do it with enviable efficiency.

THE AMERICAN DILEMMA

American agriculture lags Japanese agriculture in the application of high technology for the production of food. We lag for several reasons. First, and foremost, the economic incentives are not yet in place which seem to be needed before high tech comes into play. Right or wrong, and only with time will history bear this out, over the last 30-40 years American industries, including agriculture, have been rather short sighted. Long range planning has not been one of our strong points.

As a matter of fact, and this brings us to the second reason, in our great abundance wrought by overproduction we have not been forced, like the Japanese, to utilize every morsel produced. Our overproduction is wrought not by efficiency but by brute force - big acreage, big machines, powerful chemicals, etc. - and in the process we waste more than any other nation.

To change this trend will not be easy. Some say it will take an agricultural disaster to make us become efficiency minded. I hope we have not become calloused in our riches. The incentives for becoming more efficient in our abundance are clear. The third world wants our overabundance; they hate us for our wasteful ways. We impress the world by our high technology; we fail the world by not using this technology to make agricultural production more efficient.

In the early 70's I conducted a telephone survey of the major food stores in North Carolina in order to ascertain the "shrinkage" of fresh foods reaching retail outlets. Estimates ranged from 5% to 11%. While some portion of these losses must be attributed to improper handling, packaging, and shipping I submit that a significant portion of the shrinkage must be related to poor or lack of sorting for quality and/or chemical composition. If only 2-3% of the shrinkage is attributable to poor sorting methods, this amounts to several millions of dollars lost annually in North Carolina alone. Whatever the reasons a significant portion of the "food potential" is lost. The food

The author is: W.F. MCCLURE, Professor, Biological and Agricultural Engineering Department, North Carolina State University.

crisis is still with us. It is imperative that we as agricultural engineers concern ourselves with developing systems which will make full use of the food potential.

Most of us will agree that when fruits or vegetables are excised from the plant the process of senescence begins which will ultimately render them useless as a food. Couple this with the fact that a mechanical harvester, being largely nonselective, gathers a spectrum of sizes, shapes, and ripeness. Therefore, this suggests that the earlier sorting can be performed the closer we will come to utilizing the full food potential. Let me illustrate. Mechanically harvesting of blueberries is performed by shaking the bush. This action produces a mixture of green, ripe and over-ripe fruit. This mixture is unacceptable for fresh markets and if left unsorted the over-ripe fraction would spoil before reaching the consumer. Early sorting removes the green and over-ripe fraction at a point in time when they still have food potential for pop-tarts, jellys, etc. Left unsorted that potential is lost by decay.

The need to reduce shrinkage and the potential for improving the quality of fresh market produce was the rationale for establishing a research project, at North Carolina State University, devoted solely to the development of instrumentation for measuring the quality and composition of agricultural products. Our early work was devoted to two areas of technology: (a) study of the interaction of electromagnetic energy with agricultural products (i.e. optical properties) and (b) the utilization of the information gained in these studies to design systems for sorting and grading. The M-belt Sorter (2) is an example of a machine which places at the hands of the grower the technology for separating blueberries according to ripeness - greatly enhancing his ability to compete in the fresh market while giving him the tools to make full use of the less desirable green and over-ripe fraction - thus, taking full advantage of the food potential of his blueberry crop.

ROBOTICS IN JAPAN

In 1980 I was invited by the Japan Society for the Promotion of Science to spend two months in Kyushu University working with faculty and graduate students who were developing machines for sorting agricultural products. Professor Yutaka Chuma, Head of the Department of Process Engineering, was my host. Collaborative studies were started two years earlier when he visited North Carolina State University and we continued that work while I was there (9, 10, 15) and brought much of it to fruition.

I visited the research facilities of several companies including Hitachi, Mitsubishi, Sony and Toyota. I was surprised to find Hitachi Electronics financially supporting a "full blown" research program in biotechnology, a program designed to maximize food production from plant-environmental systems. This program was being conducted in a multistory facility similar to the Russell Research Laboratory in Athens, Georgia. Each research team was made up of several Ph.D.'s, two or three technicians, and impressive support equipment which was up to date. Computer control systems were abundant.

I was not surprised but still amazed to observe automation in the Toyota manufacturing plants. It was truly a new experience for me to see engine blocks going through the boring and honing operations untouched by human hands. I was less surprised but fascinated by the computer vision systems employed in their microelectronic manufacturing operations.

Kyushu University, in cooperation with Japanese industries, was working on several sophisticated electronic systems for sorting fruits. One team was developing reflectance technology for sorting pears and oranges. To be ultimately controlled by a computer system, the sorter was to be designed to detect both surface defects and maturity.

Mr. Ohura, under the direction of Professor Chuma, was investigating delayed light emission as a technique for measuring maturity. At first I was skeptical of this technique but later changed my mind when he showed me on-line data which indicated increased sensitivity to maturity and projected sorting rates of one/millisecond.

In Wakayawa (approximately 150 miles southeast of Fukuoka) personnel at the Wakayawa Fruit Research Station were feverishly developing sorting systems employing computer vision. Digital cameras operating under control of a 16-bit word size computer recorded information relative to both maturity and surface defects as the fruit passed through a highly illuminated area. Computer programs provided the researcher with tools to evaluate the various views of the product. Both positive and reverse video could be studied in an effort to extract effective patterns. Pusher mechanisms and air blast ejectors were being investigated for removing rejected product from the flow. I later saw a farmers cooperative facility using this technology for sorting oranges. They were also investigating "sorting-in-flight" technology where the fruit falls in free flight through the viewing area and multi-camera systems make judgements and sorting is effected before flight is completed.

CUCUMBER ROBOT

My interest in computer vision systems perked up on my visit to Miyazaki University. My host, Dr. Yoshiichi Okada took me to see a cucumber sorting facility which was part of a farmer owned cooperative. It was this facility which convinced me that American agricultural engineers were behind in applications of computer vision in agriculture; it also convinced me that we needed to move quickly in North Carolina to become involved in this kind of research.

The system (see Figure 1) was built by Mitsubishi Electric Corporation (11, 12, 13, 14, 16, 17) and purchased by the farmer cooperative for sorting cucumbers into nine categories (Figure 3). Referring to Figure 1 and Figure 2, the cucumbers were placed in buckets or pans by hand on a conveyor, one cucumber per pan. A single line scan camera positioned above the sorting line inspects the three lines of pans as they pass under it and sends shape-pattern information to the microcomputer based sort-control unit.

On the basis of these signals (7,8), the sort-control unit grades and classifies the cucumbers and relays the result to a separate microprocessor based drop-out (sorting) control unit. The function of the sorting control unit is to drop the pans downward at the required cross-conveyor. At the same time the drop-out unit signals the display counter and the line printer to register the result of the sorting decision for each cucumber.

The size of the pans (4,5) accommodates most cucumbers that fall within the proper grade range. Any part which pertrudes beyond the edge of a pan precludes measurements, an alarm is sounded, and the cucumber is rejected into a "miss" classification. The pan is bright white in order to contrast with green cucumbers.

The camera consists of a photodiode array with the elements set in a linear array (6). The number of bits (or pixels) may be preselected for the application (265, 512, and 1024 pixels). A 1024 bit array is used for cucumbers.

The grading scheme is shown in Figure 3. Cucumbers are classed by length as 2L, L, M, S and D in descending order. Irregular shapes are given the grade A, B, and C in descending order for each of the three factors of curvature, thickness, and thickness differential respectively. Assessment of grades for irregularity of shape usually involves a length attribute - for example A2L and AS. However, grade C stands by itself (i.e. without regard to length).

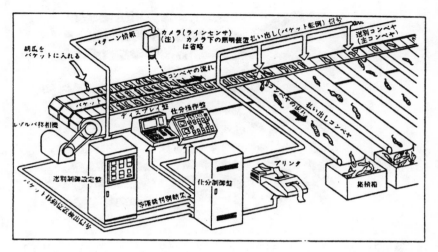

Fig. 1. Overall Schematic of the Mitsubishi Cucumber Sorting Robot.

Fig. 2. Loading Area and Viewing Station of the Cucumber Sorter.

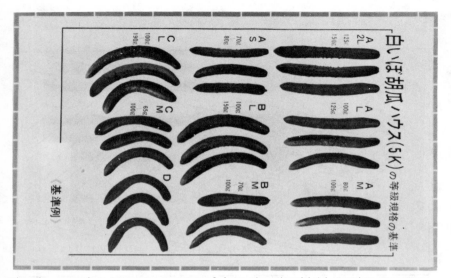

Fig. 3. Cucumber Patterns Programmed into the Mitsubishi Sorting Computer.

Problems encountered in pattern recognition include signal noise caused by stains on the sorting pan, abnormal reflections caused by the unevenness of the cucumber surface, dirt, and drops of water and other foreign matter adhering to the cucumber. Manual loading, as was done in this case, eliminated much of the contamination problem while leaving the more critical decisions of pattern recognition and accessment to the computer. Flowers and stems attached to the ends of the cucumber were "erased" by the pattern algorithm (i.e. was ignored).

Each sorting line, three pans wide, was controlled by two 8-bit word microprocessors with 12K bytes of memory. There were two sorting lines in the facility in Miyazaki and each line cost the cooperative approximately $120,000 each. Loading and packing was done manually. Crooked cucumbers were straightened and packed tightly in the top of boxes of straight cucumbers so that by the time they reached the fresh market they would remain straightened on the shelf, a rather ingenious approach to the food potential concept.

                              THE CHALLENGE

So we begin to visualize the challenge. The challenge is not to become involved in high technology simply for the prestige of being in the technological forefront. The challenge is to be involved because it will help us take full advantage of our food potential, provide us with a higher quality fresh product, and will speed the flow of food to the consumer. Robotics in agriculture will grow up in a structured environment - in the packing shed, in the processing plant, etc. - as it has done in other industries. The initial systems will be of the non-humanoid form like the Japanese cucumber sorter and the M-Belt sorter.

The Instrumentation Research Laboratory at North Carolina State University is committed to computer vision research. Two projects are currently under consideration: (1) grading sweet potatoes, and (2) egg candling. We plan to approach the work in three phases: (1) computer vision research and the development of vision software, (2) develop tactile sensors which are vitally needed before true humanoid robotics can be applied in agriculture, and (3) robot design (both humanoid and non-humanoid types). We trust that others will take up the challenge as we have so that when our day comes, U.S. agricultural engineers, as the Japanese, will be in a good position technologically to take full advantage of our food potential.

REFERENCES

1. Chuma, Yutaka, K. Nakaji and W.F. McClure. 1982. Delayed light emission as a means of automatic color sorting of persimmon fruits (Part 1): DLE fundamental characteristics of persimmon fruits. J. Fac. Agr. (Kyushu Univ.) 27(1):1-12. (English).

2. Chuma, Yutaka, Kei Nakaji, and W.F. McClure. 1982. Delayed light emission as a means of automatic color sorting of persimmon fruits (Part 2): DLE characteristics as a means of color sorting. J. Fac. Agr. (Kyushu Univ.) 27(1):13-20. (English).

3. Chuma, Yutaka, K. Morita, and W.F. McClure. 1981. Application of light reflectance properties of Satsuma oranges to automatic grading in the packinghouse line: relationship between grading index and spectral reflectance. J. Fac. Agr. (Kyushu Univ.) 26(1):45-55. (English).

4. Doi, Junta. 1976. On-line processing of agricultural crops by pattern recognition. J. Soc. Ag. Mach. 38(3):353-358. (Japanese).

5. Doi, Junta. 1976. Discrimination of agricultural crops by infrared pattern processing. J. Soc. of Ag. Machinery 39(2):151-156. (Japanese).

6. Ikeda, Y., R. Yamashita, and Y. Matsuo. 1980. On the system evaluating the shape of farm products via image processing technique (Part 1): Research Report on Agricultural Machinery 10:80-91. (Japanese).

7. Ikeda, Y. and Y. Matsuo. 1981. Selection systems for agricultural products. Proceedings of the 40th Meeting of the Soc. of Ag. Mach., p. 158. (Japanese).

8. Ikeda, Y., R. Yamashita, and Y. Matsuo. 1982. Pattern recognition research: Proceedings of the 41st Meeting of the Soc. of Ag. Machinery, p. 202. (Japanese).

9. McClure, W.F., R.P. Rohrbach, L.J. Kushman, and W.E. Ballinger. 1975. Design of a high-speed fiber-optic blueberry sorter. Trans. of the ASAE 18(3):487-490.

10. McClure, W.F. and R.P. Rohrbach. 1978. Asynchronous sensing for sorting small fruit. Agricultural Engineering 59(6):13-14.

11. Nomura, Y. 1979. Automatic cucumber selector. Mitsubishi Elec. Tech. Report 5:13-19. (Japanese).

12. Nomura, Y., O. Ito, M. Naemura. 1979. Development and application of the Mitsubishi pattern recognition/selection system (MELSORT). Mitsubishi Elec. Tech. Report 53(12):899-902. (Japanese).

13. Nomura, Y. 1980. Mitsubishi automatic sorting system. Special staff report. Mitsubishi Electric Corporation, Tokyo, Japan (November). (English).

14. Nomura, Y. 1981. Application of the Mitsubishi Elec. Pattern Recognition System. J. Image Information 11(12):65-70. (Japanese).

15. Rohrbach, R.P. and W.F. McClure. 1978. A production capacity conveyor for small fruit sorting: The M-Belt. Trans. of the ASAE 21(6):1092-1095.

16. Shimatachi, Y., Y. Nomura, I. Ide, and O. Ito. 1982. Application of pattern recognition technology in the fish industry. Mitsubishi Elec. Tech. Report 56(3):242-246. (Japanese).

17. Takizaua, T., M. Hirabayashi, I. Handa, F. Mukai, H. Nagato. 1982. Applications of the Mitsubishi pattern recognition and selection system. Mitsubishi Elec. Tech. Report 56(10):778-782. (Japanese).

Table 1. Specifications of the Mitsubishi Automatic Sorting System for Sorting Cucumbers

Feature	Specification
1. Maximum processing capacity	36,000 pcs/hr
2. Conveyor speed	23 m/sec
3. Field of view (one pixel)	1.2 x 1.6 mm
4. Sorting standards	Length, oddness[a]
5. Maximum measurement range	Length: 260 mm
	Thickness: 40 mm
	Curvature: 60 mm[b]
	Max Thickness:30 mm

[a]Oddness includes such features as thickness (diameters), curvature, maximum thickness differential.

[b]Maximum difference between length of external and internal curves.

Table 2. Features and Functions of the Mitsubishi Cucumber Sorter

Feature	Function
1. Sorting bucket	Provides background for camera
2. Camera	A single array camera scans three buckets at a time (arrays of 256, 512, and 1024 pixels may be chosen)
3. Faulty placement detection	Cucumbers overlapping the bucket will give an error signal and rejected at the "miss-line"
4. Automatic chain stretch correction	The chain stretch with time is compensated by a unique monitoring system
5. Printer	A printout is provided showing total units sorted, number within each grade, and price
6. Input console	A console located at the conveyor loading station minimizes time between farmer lots
7. Other products	Apples, pears, melons, and fish are sorted by projected area

Table 3. Components of the Computer Vision System in the Instrumentation
          Laboratory, Department of Biological and Agricultural Engineering,
          North Carolina State University

Component	Function
Intel Series III Development System with 1 M-byte of memory, DMA controller, color display, 22 M-byte disk	Camera control and data acquisition, data processing
Reticon Model MC520Y Matrix Camera (100x100 pixels) Reticon Model LC110 Line Scan Camera (256 pixels) Reticon Model RS520-08 Controller	

IMAGE CONTROLLED ROBOTICS IN AGRICULTURAL ENVIRONMENTS

E. G. TUTLE

Citrus fruit harvesting has reached technical maturity using manual, mechanical, and chemical methods. Robotic machines are approaching practicality by using electro-optic techniques that are adopted from the aerospace industry. Using these newer technologies requires a 'systems' analysis of the entire harvesting process from tree layout through transport from the grove. Such an analysis enabled concepts to be presented here and is based on orange, grapefruit, and lemon fruit-growing and harvesting realities.

Martin Marietta's Opto-Robotic™ Systems benign harvester uses image process controlled robots, selectively severs fruit, and delivers it to a collection system. A typical configuration consists of three, semiautonomous robotic modules that can work one half a tree and then, without repositioning, work the opposite one-half tree in the adjacent row. Each module contains the optical, and robotic components, image processing and control computers, and harvest zone illumination. This novel design uses two different optical systems and will harvest at night.

THE INDUSTRY AND ENVIRONMENT

In the United States, approximately 1,200,000 acres are devoted to citrus production, with over 750,000 acres in Florida. California is second in acreage followed by Texas and Arizona. The crop is divided into process and fresh fruit, the latter is sold as whole fruit and one- quarter of the U.S. crop is for fresh fruit mostly produced in California and Arizona. Only 7 percent of the Florida crop is for fresh fruit. Recent on-tree value of U.S. oranges was in excess of $1,055,000,000 with the Florida crop valued at $820,545,000 produced from over 46,000,000 trees. The world wide production of oranges is about 31,000,000 metric tons with the U.S. producing about 8,725,000 tons. Mechanization of harvesting becomes important to enable competitive pricing in the world market. The fresh fruit crop is particularly labor intensive to enable harvesting with minimal damage. Less damage allows longer storage life during distribution.

Opto-Robotic™ Systems is a trademark of Martin Marietta Corporation, U.S.A.

Mr. E. G. Tutle is Manager, Technology Evaluation and Transfer, Martin Marietta Orlando Aerospace, Orlando, Florida 32855, U.S.A.

Process fruit crops are not as demanding as they are used quickly and are harvested using techniques that can be more deleterious to trees.

Manual harvesting is slow and hazardous particularly in the upper parts of the tree. Damage to trees occurs and working height limits are 25 feet. One worker can typically harvest ten 90-lb. boxes in one hour and average Florida trees yield between 6 to 10 boxes. The average worker's day is 6 hours and the harvesting effort can be great as payment is made on the number of boxes produced. The availability of workers, however, is decreasing, replacements are needed to make up losses to other industries, and there are other demographic reasons. Machines, in addition to manual efforts, are now a necessity to produce fruit profitably and in a timely manner.

New harvesting techniques are particularly attractive for premium crops if such techniques improve productivity and quality. Selective harvesting, using robotic means, and operating for longer periods at night will enable growers to meet these goals. Reliable equipment specially designed for the environment is now becoming available.

CONDITIONING THE ENVIRONMENT

Robotics requires disciplines of design and of the work place. The nature of trees and fruit establish operating rules, but the work place might be modified to optimize the design and process. Figure 1 suggests possible 'standards' for trees planted on 25 or 30 ft. grid centers, which is typical in the U.S. These dimensions are for mature trees and provide space for practical harvesters working between rows. Hedging the path-sides of each tree is a current practice that appears to increase production and is recommended.

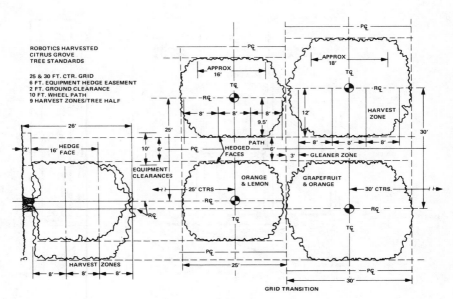

Fig. 1   Tree/Grove Standards

For practical optical and mechanical reasons, a preferred working zone in a tree was determined to be approximately 8 x 8 x 12 feet (H, W, D). This is compatible with current U.S. tree/grove management practices and suggests a standard approach to harvesting. Each half of a tree is divided into nine harvest zones of equal size, three across and three high. The reference for harvester rig centering is the tree trunk.

Presently, trees on 30-foot centers would not be completely harvested because of possible 'gleaner' zones. These could be harvested by partially moving the rig to center operations at the midpoint between trees, but this may not be cost effective.

For technical reasons, trees planted on 24-foot centers (or the nearest practical metric equivalent) in the row would be preferred. With harvester designs to make it practical, taller trees can now be encouraged. We also recommend tree skirt clearances that would provide fruit-laden tree ground clearances of two feet. The harvest rig will require a broad wheelbase and outriggers for stability, particularly for groves with tall trees.

THE HARVESTER

The tree, grove, and harvester design has been rationalized in the form of a three-module machine (Fig. 2). The modules accommodate the recommended harvest zones and can be moved easily through the groves. The criteria for operation in its intended environment is based on practical sizing. Assume for the moment that a practical harvest module is available.

The modules harvest the trees from the top down to allow 'unloading' the tree in that manner. As the branches are unloaded, more fruit is exposed and the remaining fruit moves upward, sometimes into the space previously occupied by

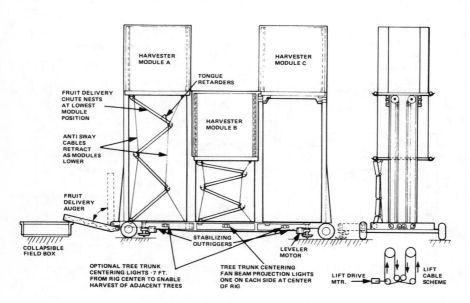

Fig. 2   Three-Module Harvester

the fruit just harvested. Each module is elevated to the highest of three levels for the size of tree to be harvested. As each zone is harvested, the module is programmed to descend or stop. To work the opposite tree row, the center module is raised to the highest level and rotated 180°, then lowered to the middle position. The outer modules are then raised and rotated and the harvest sequence begins again. The center module, lowered to allow the outer modules to rotate, is now raised in proper sequence and begins its harvest.

The harvester has its own fruit delivery system starting with troughs on each side of the respective modules that contain motor driven wire augers. Fruit harvested in each module is urged to the gravity feed troughs which direct the fruit to troughs in the harvester base. These troughs have wire augers that drive the fruit to a pair of lift augers that direct the fruit into collapsible collection boxes. These boxes are designed to be placed between the trees and will clear the harvester as it moves through the grove.

The harvester is positioned so the center module is opposite the tree trunk. (The prime mover is a field tractor that also contains an electric power generator and appropriate controls for the harvester.) High intensity projection lamps on each side of the harvester base direct vertical fan beams that act as cursors to find the tree trunk. A field worker follows the harvester and signals the tractor driver when the light intercepts the tree trunk. The driver stops and sets the wheel brakes. An automatic sequence starts that swings out stabilizing outriggers. Leveling motors at the ends of the outriggers are directed to anchor and level the harvester. The field worker meanwhile erects the now exposed, collapsed collection box and positions it under the egress of the lifting auger that is pivoted at the rear of the harvester. The worker now starts the delivery system and the harvester is ready.

Since the harvest process can be automatic, once the harvester and collection boxes are positioned, the driver and field worker would be free until the harvest is complete. One pair of workers could readily work three harvesters placed on non-adjacent paths. Harvesters should not work opposite sides of the same tree at the same time, for reasons which will be obvious later.

The Harvester Module and the Harvest Zone

Viewing the harvest zone is done with a global sensor that contains a 256^2 photo diode array behind an appropriate lens and filters. Its first position is at the geometric center of the module and at the rear of the module opposite the harvest zone. The optical system is arranged so the view through the aperture is focused on the entire square photo-diode array, which is one reason for the square harvest zone.

The view into the harvest zone is shown in Fig. 3. The lower and center zones of a large tree will appear to cover the aperture. The upper and outer zones will show fruit near the aperture and at various distances. Some fruit will be in clusters and some will be partially obscured by leaves or branches (1). Further, some fruit can be framed by sky. In the daytime, lighting conditions are variable and some fruit will be in a shadow. The difference in light levels makes it difficult to locate the fruit. Also it has been found (2) that light reflecting from leaves in the sun is four times brighter than fruit in the shade. To negate these problems requires harvesting at night. This radical departure from convention has numerous advantages, one of which is the longer (and cooler) working 'day' which can be from twilight to shortly after dawn. Better control of lighting conditions also enables using selective filters that are critical to fruit-leaf discrimination, discussed later.

The robotic mechanism is required to harvest all the locatable fruit that appears across the aperture. In this instance, the practical fruit depth is about 1 meter into the canopy. In the upper and outer zones, the canopy depth is the same, but the fruit distance may be to the center line of the tree row.

The problem is to locate all the fruit that can be observed from one point. Tungsten flood lamps inside the module are turned on and, for a short time, light up the harvest zone. The scene is captured by the global sensor and stored in an electronic memory. The scene is illuminated with a constant light level and from several points, which tends to negate shadows. The fruit appears in contrast to the leaves. One sensor and multiple light sources are used as the sensing system and only one system is needed in each module. Fruit location by triangulation is not possible as there is no assurance a specific fruit can be viewed simultaneously by two or more sensors. However, light reflected from observable fruit travels a path through an imaginary screen at the module aperture to the sensor. This imaginary screen is imaged on the 256^2 photo diode array with specific fruit subtending three or more diodes on the array. A 7 cm diameter orange at 6 meters from the sensor will be imaged over a cluster of at least 5 diodes. Nearer fruit of the same size will image over a larger array of diodes.

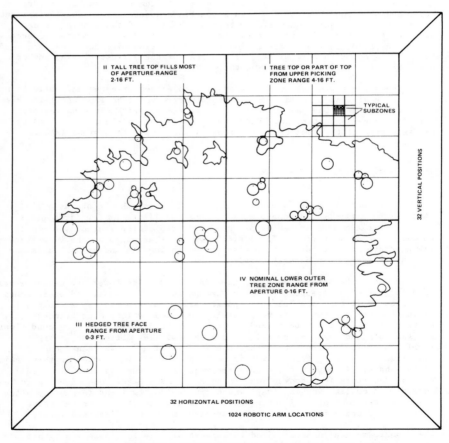

Fig. 3 Harvester Aperture Views with Subzone Structure

If the aperture were divided into 1024 subzones, each subzone would subtend 64 diodes of the 65,536 that comprise the global sensor array. The cluster of 64 is shown hypothetically, as if it were at the aperture. As our interest is only to detect sufficient reflected light in certain spectra, this resolution of the zone scene is adequate.

As a 7 cm diameter orange essentially subtends one subzone at the aperture, the robotic arm needs to be directed to any one of 1024 subzones. Thus the directional coordinates for the robotic mechanism need not exceed 1024.

The above resolutions are also important since the image processed fruit ultimately appear as points within the subzones. The robotic machine needs only to be directed into subzones showing fruit points, which reduces harvest time.

The Robotic Harvester

The X-Y coordinates of the fruit, with respect to the aperture and the optical axis and vertex, are stored in memory. However, the distance to the fruit is neither known nor easily determined. This harvester, situated within the module as shown in Fig. 4, uses a novel concept to find the fruit. To positively direct the telescoping robotic arm to the right fruit coordinates, the pivot center of the arm is placed at the optic vertex; the global sensor is displaced upward as it is no longer required. The robotic arm can now traverse the 'boresight' path of the optical rays between the fruit and the global sensor.

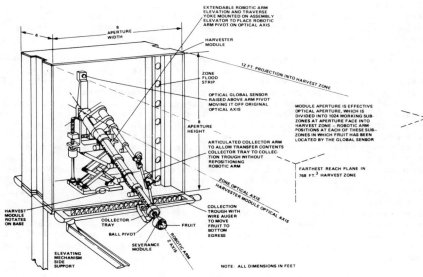

Fig. 4 Harvester Module

A second sensor, in the severance module (Fig. 5), assists in finding the fruit. A self-contained light source projects light on the optical axis of this second or 'seeker' sensor. The robotic arm with the severance module is directed by the computer to the first subzone known to have one or more fruit.

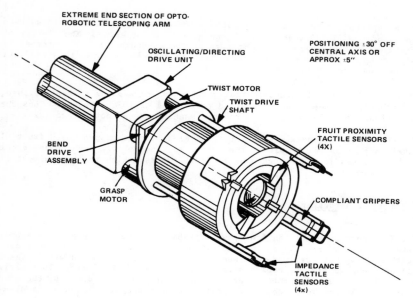

Fig. 5 Severance Module

The arm comes out of the module, past the aperture in the direction of the target fruit. The extension is rapid, on the order of 4 meters/second. The seeker light source illuminates the path of the severance module. When sufficient light is reflected from any object in its path, this event signals immediate deceleration of the arm.

The severance module is oscillating about the robotic arm axis, with its sensor scanning the path. The sensor element is a silicon quad-element, biased, photovoltaic detector (Fig. 6). A lens and spectral filter direct reflected light to the segments of the detector. Any light energy above a certain threshold triggers the arm deceleration circuit. Fruit having significantly greater reflectance than leaves, will cause one or more detector quadrants to receive more light, which stops the oscillatory motion. The seeker action will cause the severance module to be positioned to align itself so equal light energy will fall on each quadrant. The severance module, now decelerated, but still moving, scoops in the fruit as it exerts pressure on at least three of the four tactile sensors projecting into the seeker optical path. Sufficient tactile force stops further arm advancement and immediately triggers a grasp drive that causes elastomer tipped fingers to grasp the fruit. This action is immediately followed by a bending of the severance module, which begins exerting force on the stem. Then in quick succession, the severance module grasp-assembly rotates, twisting the fruit stem, and the robotic arm retracts quickly, about 30 cm. Severance can occur at any time during the bend, twist, or jerk motions. Details of various severance sequences can be found in Fig. 7.

Since the seeking severance unit contains a light source, a similar light source working the other side of the same tree could cause unwanted destructive consequences.

The robotic arm is programmed to harvest from the top down starting at the upper left, moving right, then into the next lower row of subzones, moving right to left, etc. By computer control, the robotic arm is directed only to subzones having fruit, but the sequence is in this order.

90

A secondary arm is coupled to the robotic arm, which initially follows the extension of the main arm. It positions a collection tray below the severance module and at the point of withdrawal after severance. The fruit drops into the tray. When several fruit are collected, this arm retracts to position the tray along the collection trough to discharge the fruit. Meanwhile, the main arm continues to seek, acquire, capture, and sever fruit. It stops only to await return of the tray.

The use of the collection system can be optional; the fruit could be dropped to the ground from the severance module. However, this does damage the fruit, (2). Therefore, the secondary arm and collection system are considered important, particularly in harvesting for the fresh fruit market.

Spectra and Image Processing

The operation of this harvester depends on the reflectance spectra from fruit and leaves, (2) (3). The curves (Fig. 8) show a significant ratio of fruit to leaf reflectance in the 600 to 700 nanometer spectrum. Green fruit or fruit with serious surface defects reflect less light and can be selectively rejected. Tungsten illumination at 2500°K further enhances the fruit reflectance, although lower color temperatures can be acceptable. The photo sensors also should peak in this spectrum. The Martin Marietta series LD230 quad detectors together with a 600 to 700 nm band-pass filter combined with a broadband glass lens meet our requirements. For oranges, the tungsten source in the seeker section of the severance module provides the correct color temperature illumination.

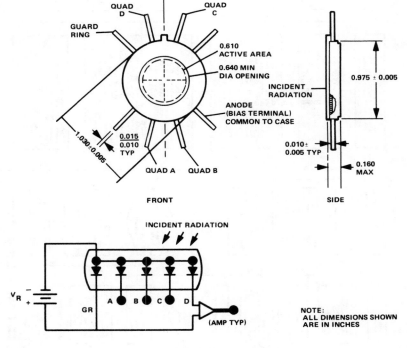

Fig. 6 Quad Detector

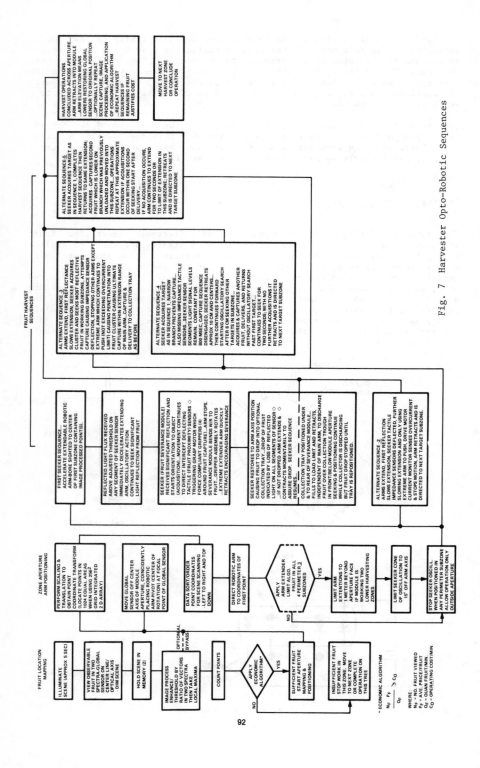

Fig. 7 Harvester Opto-Robotic Sequences

92

The global sensor advantageously uses the 650 to 750 nm peak and 900 nm roll-off of arrays such as the EG&G Recticon Corp. series RA256X256A. This is particularly important since it is necessary to compare the 600 to 700 nm and 750 to 850 nm spectra of the harvest zone reflectors. This is done by first viewing, through a band-pass filter, the first spectrum, called 'A,' and storing this information in an 'A' memory. The second or 'B' spectrum is stored in memory 'B.'

This comparison of spectra is necessary to locate fruit versus leaves when they are in the near field. This situation is analogous to a radar problem where foreground clutter signals are greater than a far field signal. Reflected energy is proportional to $1/R^4$ (where R is the distance to the reflecting object). A diode (pixel receptor) will receive 81 times more energy from a reflector at 2 meters than from a similar object at 6 meters.

The image enhancement/thresholding process, facilitated in the harvester module on-board computer, computes the ratio of reflectances (A/B). The curves in Fig. 8 show that fruit reflects 10 times more light energy than leaves in spectrum A and about two times more in B. Energy from leaves in B will be four times greater than from spectrum A. Therefore, the fruit reflectance ratio A/B = 10/2 = 5; the leaf reflectance ratio A/B = 1/4 = 0.25. Comparing these ratios, pixel by pixel, we find that the fruit amplitude is enhanced 20 times over that of leaves. Reflected clutter from leaves, particularly 2 to 3 meters from the detector, therefore, can be of a lower signal amplitude than reflectance from fruit at six meters. The threshold for later signal processing is now set to enable processing only fruit reflectance amplitudes.

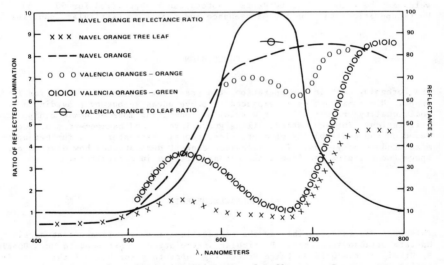

Fig. 8   Reflectance and Ratio-of-Reflectance Curves

The image processing is done as follows. The scene in each spectra is captured by the photo array and each column of 256 pixels is read out, serial fashion as a string of variable amplitude signals which are immediately processed into 8-bit digital signals by an A/D converter. This provides 64 grey levels representing reflected energy magnitudes to be stored in dynamic 65K byte, microprocessor controlled memory: the A spectrum into memory A, the B spectrum into B. A typical microprocessor is a Zilog, Z8000 series CPU and associated support elements. The output vectors are compared and the ratio computed, after eliminating all ratios below a certain threshold. This data is stored in another 65K byte memory. Then the latter memory vectors are analyzed for local maxima using a 5 x 5 sparce operator. A further understanding of image processing may be found in Pratt (4) and for microprocessors in Zilog publications (5).

The result of using the operator to find local maximas is a series of 'points' representing fruit with their coordinates available as 8 bits in X and 8 bits in Y. Optionally, the points can be counted to apply the Economic Algorithm, shown in Fig. 7. Coincidently, all possible points (65,536) are now divided into 1024 zones ($265^2/64$) and the coordinates of zones that contain one or more points are stored in a smaller memory. These zones are then reordered to be listed in the order the robotic arm will traverse and elevate or depress. Thus, the arm will be directed only into subzones known to contain fruit. The program-controlled microprocessor calls up the first coordinate to position the arm. After completing the subzone operation, a new set of coordinates are called up that redirect the arm to a new subzone, etc. Servo control and other operations in the severance module are controlled by another, smaller, microprocessor in that module; it is subordinate to the main microprocessor.

A third microprocessor, part of the harvester assembly, controls positioning, leveling, harvester module elevation and rotation, and cooperatively controls the computers in the harvester modules. All control is remotely centered in an enclosure located on the field tractor, where certain field options and overrides may be entered. The computers are designed to operate in field environments. All control is independent of generator operating frequencies, which may be either 50 or 60 Hertz. Motors can be specified for 50 or 60 Hertz operations to simplify maintenance and repair in particular countries.

CONCLUSION

The foregoing details the operation of a practical robotic harvester for fruit crops. Night harvesting is expected to prove advantageous by providing a cool, quality product. Longer working periods allow more economic equipment utilization. Further, standardizing grove layouts and management practices, based on the availability of a practical robotic harvester will further optimize production yields. This harvester concept demonstrates how diverse aerospace technologies can find practical application in agriculture.

ACKNOWLEDGEMENTS

This work was made possible by the cooperation of many colleagues at Martin Marieta Orlando Aerospace. Particular appreciation is extended to Mr. Donald R. Ziesig of the Orlando Image Processing Laboratory for his valuable consultation and suggestions. The author is particularly grateful to Drs. Isaacs, Shaw, and Whitney and Mr. Glen Coppock of the University of Florida, Institute of Food and Agricultural Sciences for their encouragement and cooperation in providing vital data and information about citrus agriculture. Patents are pending for the harvester system described in this paper.

(1) C. E. Schertz and G. K. Brown. 1966. "Determining Fruit-Bearing Zones in Citrus." TRANS. of the ASAE: 366-368.

(2) C. E. Schertz and G. K. Brown. 1968. "Basic Considerations in
     Mechanizing Citrus Harvest." TRANS. of the ASAE: 343-346.

(3) J. J. Gaffney. 1973. "Reflectance Properties of Citrus Fruits." TRANS.
     of the ASAE: 310-314.

(4) W. K. Pratt. 1978. "Digital Image Processing." J. Wiley and Sons

(5) Zilog Microprocessor Applications, Ref. Book, Vol. 1, Z8000, Sect. 3,
     Zilog Corp., Cupertino, CA, 95014, ©1981.

NUCLEAR MAGNETIC RESONANCE IMAGE INTERPRETATION

M. W. Siegel

Introduction

Nuclear magnetic resonance (NMR) is an analytical tool used in physics, chemistry, and biochemistry. It has recently been adapted to the generation of images, primarily for medical diagnosis[1]. The information is inherently three dimensional, i.e., the instrument reports the spatial density of specific susceptible nuclei within its volumetric field of view. This information may be displayed in three dimensions, or as a stack of two dimensional sections. The power of NMR imaging in medical diagnosis is its unique ability (e.g., as compared to x-rays) to provide images with strong contrast between soft tissues of similar density but different physical structure. The technique is regarded as non-invasive, in that no ionizing radiation is involved, and so far as is known it causes no biological damage.

In view of its proven medical imaging applications[2], it is interesting to speculate on the possibility of applying NMR imaging to other fields, e.g., agricultural automation. Its ability to non-invasively provide three dimensional images of the interior of solid objects means that in principle it is possible, for example, to find oranges hidden from view by limbs and leaves, and simultaneously to judge their ripeness on the basis of chemical and structural characteristics related to maturity. Present technology does not permit this application: NMR instruments are, to put it mildly, not portable. On the other hand, existing instruments are in principle suitable for fixed location applications, for example, inspection and sorting. Unfortunately their expense, complexity, and slowness (typical imaging times are presently several minutes) now make even fixed location applications impractical except for very high unit value specimens, e.g., people, components of spacecraft and nuclear reactors, etc. This discouraging scenario notwithstanding, there are in principle no insurmountable obstacles to making NMR imaging economically viable in fixed location applications, and it is reasonable to expect that future engineering breakthroughs will make it technically and economically viable in mobile applications as well.

This paper presents background information on analytical and imaging NMR techniques, discusses image interpretation in terms of its image processing and image understanding aspects, and illustrates present NMR imaging and standard image interpretation technologies by showing the application of several image interpretation software packages, developed for analysis of aerial photographs, to NMR brain scan images[3].

The author is: M.W. SIEGEL, Senior Research Scientist, The Robotics Institute, Carnegie-Mellon University. This paper is based in part on a presentation by Raj Reddy and M.W. Siegel at the Second C-MU Conference on Biological Spectroscopy, San Jose, CA, February 8-11, 1983.

Principle of Nuclear Magnetic Resonance

If a nucleus has an odd number of nucleons (protons and neutrons together), or if it has an even number of nucleons formed of an odd number of protons and an odd number of neutrons, i.e., if it has either an odd atomic weight, or an even atomic weight and an odd atomic number, then (and only then) it has a magnetic dipole moment. Because the magnitude of the magnetic moment is characteristic of the nucleus, its measurement can serve to identify and quantify the presence of these nuclei. Common nuclei of biological interest which fall in this class include hydrogen, nitrogen, sodium, and phosphorus. The major isotopes of common species such as carbon and oxygen do not fall in this class, but some of their natural isotopes, such as C^{13} and O^{17} do.

The NMR method is based on the measurement of the orientation energy of the magnetic moment in a known magnetic field. The characteristic orientation energies in the magnetic fields used are in the radio frequency range, i.e., tens to hundreds of megahertz. The sample is irradiated by a brief, intense radio frequency "chirp", which reorients some of the susceptible nuclei into higher energy orientation states. They spontaneously decay from these excited states, emitting characteristic narrow-band frequency signatures. A sensitive radio receiver detects and records the decay signal. In detail, several different excitation and detection protocols are commonly used, each of them providing differently useful kinds of information.

The fundamental frequency identifies or selects the nuclear species of interest. The intensity of the emitted signal depends on the number of emitting nuclei in the sample, so measurement of the intensity determines their spatial density. The chemical environment of the nuclei, i.e., the electrons of the molecules into which they are bound, slightly modifies the local magnetic field and thus the emitted frequencies, so detailed analysis of the frequency spectrum provides a chemical analysis, or alternatively, is used to deduce molecular structures of unknown compounds. The physical environment of the molecules , i.e., whether they are present in gas, liquid, glassy, or crystalline matrices, determines (in part) the rate of their spontaneous decay from higher energy to lower energy orientation, so measuring the decay rate of the emitted signal characterizes this environment.

While this overview is necessarily simplified, it does convey the fundamental principle: the radio frequency emission spectrum of susceptible nuclei in a known magnetic field can be used to quantify their concentration, chemical environment, and physical environment. The physical environment, characterized by decay rate (or its reciprocal, relaxation time), is of major interest in imaging applications, since it may be used to provide contrast between tissues of similar density but different structure.

Principle of Nuclear Magnetic Resonance Imaging

For practical reasons, NMR imaging is presently restricted to hydrogen nuclei. Since biological materials are composed primarily of hydrogen (by atom count) this restriction is not very severe.

In analytical NMR, great pains are taken to assure the uniformity of the magnetic field. If the field is inhomogeneous, then identical nuclei in different regions of the sample emit different frequencies, and chemical analysis is hopelessly confounded. In imaging NMR this problem is turned to an advantage: the magnetic field is intentionally made inhomogeneous, with a precisely specified spatial gradient direction and magnitude, so the frequency at which hydrogen nuclei radiate marks their location along the gradient. The relaxation time remains characteristic of the physical environment. The three observables, frequency, intensity, and relaxation time, respectively determine where, how much, and in what environment hydrogen nuclei are located. By their analysis sectional and three dimensional images are constructed. In medicine, these images are able to segregate bone from normal soft tissue from diseased soft tissue; in future agricultural applications they might, for example, be able to segregate wood from fruit, ripe fruit from green fruit, and healthy fruit from diseased or damaged fruit.

Three logically separate steps, each requiring specialized computer software, and preferably specialized hardware, are used to reduce NMR signals to useful images. The raw data consist of a large set of frequency spectra, each spectrum corresponding to a predetermined direction of the magnetic field gradient. In the first step, the spectra are fourier analysed, transforming them into hydrogen nucleus density distributions along the corresponding gradient axes. In the second step, algorithms similar to those used in x-ray tomography transform the axial density distributions into two and three dimensional density distributions. In the third step, image processing programs suitably transform the data (as discussed below), and represent them as images on a computer output device, usually a cathode ray tube. We anticipate a future fourth step, automated image understanding, coupled with automatic control of the NMR instrument and the tasks being monitored by the instrument.

Image Interpretation

Image interpretation involves two steps: image processing, transforming the raw image data into easy to understand form, and image understanding, attaching meaning to parts of the processed image.

Modifying an image's brightness and contrast is a simple kind of image processing familiar to everyone who has adjusted the controls of a television set. Many similarly useful transformations of image data are easily accomplished when the image is stored in the memory of even a small computer and modified by simple pixel-by-pixel operations. These transformations, involving changes in the intensity scale (including the use of false color), spatial frequency content, and local topology, simply accentuate image regions in which the information density is high. They make no reference to the meaning of the image, except perhaps through the intervention of human operators who have knowledge of the actual or expected content, and use this knowledge to choose operations and parameters which produce particularly pleasing or meaningful images.

Until recently image understanding, the extraction of contextually meaningful features from image data, has been an exclusively human job. With the advent of the need to interpret large numbers of images generated by aerial and satellite photography, industrial inspection systems, medical screening programs, etc., a great deal of impressively successful effort has gone into developing computer software and hardware systems for automated image processing and understanding. These systems, which require large computers to produce their results in reasonable times, attempt to use all the available contextual knowledge both for determining the most suitable image processing steps, and for the extraction of meaningful features, i.e., understanding. They make use of information sources external to the image itself, including data and rules based on prior experience. The output of image understanding systems is high level symbolic information which, combined with additional data and rules, can be translated into control strategies and control signals for automated mechanical devices.

NMR Image Interpretation Examples

Figure 1 shows one of a set of several brain scan sections[4] reconstructed in planes perpendicular to the spine. This section is more-or-less through the eyes and ears, with the eyes and nose being clearly identifiable at the bottom of the figure. It is an "honest" representation of the raw image data, i.e., neither digital nor analog processing (adjustment of CRT monitor brightness and contrast) have been applied. In this and all subsequent images, the CRT was photographed with an exposure which was determined to be correct for midscale grey when the contrast and brightness were adjusted to reproduce black as black and white as white. The raw data span eight bits, with intensity 0 represented as black, intensity 127 represented as midscale grey, and intensity 255 represented as white. The original image contained 128 x 128 = 16384 pixels. A cubic interpolation scheme was used to increase the image size to 256 x 256 = 65536 pixels for clear visibility. Since each pixel has a range of eight bits, representing the entire image requires 524288 bits.

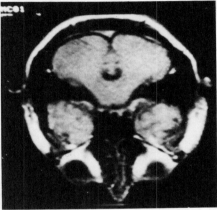

Figure 1 Raw data of an NMR brain scan image. The final photographic printing step has resulted in some (undesired) improvement in the visual quality of this image.

Histograms are a powerful tool for both manually and automatically identifying and tagging interesting features in an image. Figure 2 shows a histogram of the data represented in Figure 1. The abscissa represents intensity (0 to 255 intensity units) and the ordinate represents the number of pixels with each intensity. Each tick mark on the abscissa represents an intensity increment of 32 units. Note that very few pixels have an intensity above 96 units, i.e., the vast majority of pixels have intensities less than midscale grey; this explains the dark appearance of Figure 1. We will subsequently show that the large peak below 32 intensity units is mostly background noise, and that almost all of the interesting information in Figure 1 lies in the histogram peak between 64 and 96 intensity units.

Figure 2 Histogram of the data represented in Figure 1. The abscissa is pixel intensity (0 to 255), and the ordinate is number of pixels.

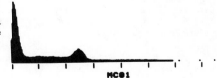

We can go a long way towards improving the appearance and usefulness of Figure 1 by digitally adjusting the contrast and brightness range, guided by the histogram of the raw data. Figure 3 shows the image which results when all pixels with intensity above 127 units (grey) are set to 255 units (white), and all pixels with intensity below 128 units are doubled in intensity. At the small price of washing out some information in the (near white) bony regions, the visibility of fine structure in the soft tissue is greatly enhanced. Figure 4 shows the histogram corresponding to the modified image.

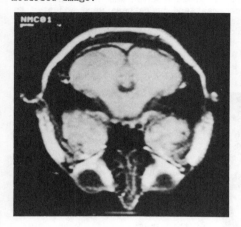

Figure 3 Normalized version of Figure 1. Intensities originally greater than 127 are set to 255, and intensities originally less than 128 are doubled.

Figure 2 Histogram of the data represented in Figure 3. The abscissa is pixel intensity (0 to 255), and the ordinate is number of pixels.

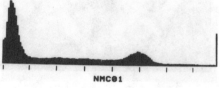

Following up on the earlier statement that almost all of the interesting (brain tissue) information is contained in the histogram peak originally between 64 and 96 intensity units, Figure 5 shows what happens when we set all intensity values below 64 units to zero (black), all intensity values above 96 units also to black, and rescale intermediate intensity values to fill the range 0 to 255. Unlike the previous illustration, in which all information was retained with adjusted brightness and contrast, this procedure, called feature extraction, discards information which is judged to be extraneous. While this procedure has clearly succeeded in separating the part of the image representing brain tissue from the background noise and non-brain tissues, doing this after the fact generates sizable quantization noise, seen as graininess due to excessive contrast. One of our hopes for the future is that realtime automated feature extraction will make it possible to adjust NMR image collection procedures on-the-fly, eliminating the quantization noise which is an unavoidable consequence of post-processing.

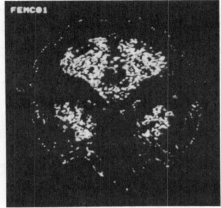

Figure 5 Feature (brain tissue)
extracted from the data by setting
all intensities not in the right
hand peak in the histograms to 0,
and rescaling the remaining
intensities to fill the available
dynamic range.

It is common in spectral methods for the most interesting features to appear
as small localized spikes ("lines" or "steps") on an intense but slowly
varying background ("continuum"). The analogy in image interpretation is
that edges or other sharp transition regions are often the most interesting
and useful parts of the image. In both cases, differentiation of the data
can serve to suppress the background and accentuate the transitions. There
are numerous numerical differentiation algorithms which have been developed
for image processing, one of which (the theta component of the Sobel
transformation, a kind of gradient operator) is illustrated in Figure 6. In
the transformed image black represents local regions where adjacent pixels
are of similar (large or small) intensity, while the greys represent local
regions in which adjacent pixels vary rapidly, whatever their base
intensity. Figure 6 clearly demonstrates that Figure 1 contains a vast
amount of information too subtle to be effectively represented by a simple
grey scale.

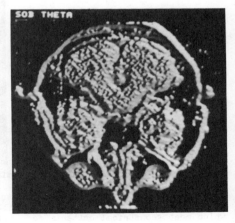

Figure 6 The theta-component of
the Sobel transformation. This edge
detection algorithm emphasizes
arc-like gradients. The left-right
asymmetry is a consequence of the
signed nature of the transformation.

Computer programs have been developed to automatically segment images into regions satisfying various similarity criteria, to mark the borders of these segments or features, and to mark the interior of those features which satisfy additional contextual criteria which tag them as being particularly interesting. Figures 7 and 8 show the application of one of these programs to the data of Figure 1. Figure 7 is an early stage of the iterative procedure; a very large number of localized regions have been delineated (in red on the original) by virtue of their internal similarity and their contrast with adjacent distinct regions. Figure 8 is a later stage of the procedure, in which similarity criteria are progressively relaxed to merge adjacent regions until only a few patently distinct features are left.

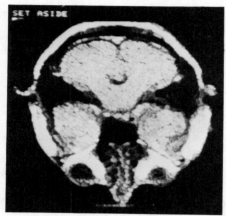

Figure 7 Output of an automatic segmentation program. The program has identified and outlined connected regions which meet intensity based similarity criteria.

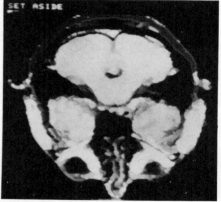

Figure 8 Further output of the segmentation program, after merging adjacent not-too-dissimilar regions. The semicircular feature at the top has been filled with color to indicate that it meets criteria for a specific type of object about which the program is knowledgeable.

In Figure 8 the interior of a thin semicircular feature has been marked (in blue on the original) to indicate that it has been tagged by the program as being noteworthy according to a set of predetermined context based geometrical cues. The features normally tagged by this program are those which meet geometrical criteria for long, thin planar objects, such as roads and rivers, which are of prime importance for understanding aerial photographs. Since the program has not (yet) been equipped with an alternative set of contextual rules which might enable it to recognize features relevant to medical diagnosis via NMR images, it has done its best to fit Figure 1 into the only context it knows: it has tagged the long, thin arc of the skull bone (in section) as the closest thing it can find to a road or a river!

[1]Pykett, I. L., NMR Imaging in Medicine, Scientific American, 24678-88 (May 1982).

[2]Kaufman, L., Crooks, L. E., Margulis, A. R., Nuclear Magnetic Resonance Imaging In Medicine (Igaku-Shoin, Tokyo, 1981).

[3]The image interpretation tools used were developed by the Image Understanding Laboratory, Department of Computer Science, Carnegie-Mellon University, and were used with the assistance of David McKeown, Ph.D., and his students and staff.

[4]The image used for illustration in this text was provided by (and is of) L. E. Crooks, M.D., Radiologic Imaging Laboratory, University of California - San Francisco. Additional images used during the oral presentation were provided by Dr. Crooks and by P. Lauterber, Ph.D., Department of Chemistry, State University of New York - Stony Brook.

INTELLIGENT ROBOT SYSTEMS:
POTENTIAL AGRICULTURAL APPLICATIONS

Jane H. Pejsa and James E. Orrock

The application of robots in the manufacturing environment is currently re-
ceiving much attention. These applications are primarily highly structured
tasks such as spray painting, arc welding and pick and place operations. In
an effort to extend the application of current robots into less structured
manufacturing environments many laboratories, research groups and universities
are investigating the integration of intelligent sensors into robotic systems.
These intelligent robot systems are aimed at tasks such as assembly and
inspection.

This paper discusses the possibility of applying intelligent robot systems to
agricultural harvesting applications. This exercise begins by delineating
those crops in the United States which have a high potential for robot applications during this decade. Based on these high potential crops, a single
task is selected for analysis. The requirements for an intelligent robot
system to perform this task are then defined. The practicability of this
intelligent robot system is evaluated today, in 1985 and in 1990.

Based on this analysis a generalization to some new insights regarding the
potential for robots in a larger sector of agriculture is made.

Potential Agricultural Applications

This analysis began by asking the question: If there is a potential for
robotic devices in the agricultural field, what agricultural comodities are
the most likely candidates? Note that this analysis was restricted to the
field environment; also, animal husbandry was excluded. To address this
question, three crop parameters were defined which say something about robot
potential. These parameters were then used to evaluate those crops for which
we had data, to define potential candidates for robot application.

The first parameter is the annual farm yield, in dollars, nationwide. The
data used was for 1979 yields for each of 43 different crops. The values vary
from $18 billion for field corn to $22 million for endive.

The second parameter is dollar yield per acre. This parameter represents crop
concentration, and the assumption is made that a more concentrated crop would
be more amenable to robotic manipulations than a dispersed crop. Among the 43
crops analyzed (again with 1979 data) values vary from more than $7000 per
acre for strawberries to $73 an acre for oats.

The third parameter is manpower requirements, perhaps the most sensitive parameter in the analysis. The assumption is made that a robotic device has the
greatest potential when applied to labor intensive tasks. Values for this

The authors are: Jane H. Pejsa, Principal Development Engineer, and James E.
Orrock, Principal Development Engineer, Honeywell Technology Strategy Center,
1700 West Highway 36, Roseville, MN 55113.

third parameter (1979 data) vary from 248 hours per acre in the tobacco industry to less than three hours per acre for rice.

Based on the assumption that commodities already automated significantly are
not good candidates for robotic application, crops that require less than 10
hours labor per acre each year were eliminated from consideration. This eliminated rice, cotton, tomatoes, peanuts and all grains from consideration,
reducing the originally considered 43 crops to 31. These remaining crops were
consequently prioritized using the three parameters just described.

First, all crops were normalized against the highest value for each parameter:
by adding together the normalized values, the 31 crops were prioritized. Five
key crops stand out as potential candidates for automation: tobacco, fruit
trees, stawberries, grapes, and lettuce. This is summarized in Table 1.
Among these candidate crops, tobacco ranks first. However, currently vast
changes in automation in the tobacco industry are taking place. Therefore,
this analysis was directed towards fruit trees, the second highest ranking
candidate. Among fruit trees, citrus trees stand out among all others. Also,
orange harvesting has been considered a task which "presently has no automated
solution" (Harmon, 1982).

Table 1. High Potential Crops for Robot Applications

	NORMALIZED PARAMETERS			
	a	b	c	a+b+c
TOBACCO	0.663	0.364	1.0	2.027
FRUIT FROM TREES 　　CITRUS FRUIT 　　APPLES 　　PEACHES, PEARS, PLUMS, PRUNES	1.0	0.205	0.560	1.765
STRAWBERRIES	0.075	1.0	0.560	1.635
GRAPES	0.345	0.205	0.560	1.110
LETTUCE	0.177	0.330	0.254	0.761

a - normalized annual farm yield
b - normalized dollar yield per acre
c - normalized manpower requirements

Based on the previous discussion and the challenge of an apparent "automation
defying" task, this exercise was continued by pursuing a potential intelligent
robot system for harvesting oranges.

Current Orange Harvesting Techniques

Before postulating an intelligent robot orange harvesting system, it is important to understand how oranges are currently harvested. In each and every
commercial orange grove, sometime between December and June, depending on
location, a sampling expert surveys the grove and examines several oranges.
When the grove is declared ready to be harvested, every orange on every tree

will be picked. A human picker equipped with a ladder and a bag enters the grove. The picker positions his ladder against the first tree; he takes stock of the distribution of oranges on the tree; climbs the ladder; then reaches out to each orange, one by one, grasping, pulling and dropping the fruit--either into his bag or on the ground below. Amazingly, the best of orange pickers can pick at the rate of an orange per second over several hours. An average picker harvests at the rate of one orange in two and a quarter seconds. According to the Florida Department of Citrus, this average picker harvests 1600 oranges per hour at piecework rates equivalent to $5.00 per hour.

An Intelligent Robot Orange Harvester

In order to address the problem of robotic orange harvesting, a system is now postulated and the performance requirements are defined in order to analyze the feasibility of this task.

Figure 1 shows the postulated robot orange picker. It consists of one or more arms mounted on a central elevator support. The system moves along a path between rows of orange trees in a grove stopping at tree locations as shown. The arm harvests from the exposed hemisphere of each tree, and then moves to the next set of trees. The harvesting at each tree occurs as follows: once the system is stationed at a tree location, the arm(s) are elevated to a pre-determined height. The height is a programmable parameter based on the average tree height in the grove. With a camera mounted on the arm, a global picture or series of pictures is taken to define the perimeter of the orange-bearing foilage. Also, with a dense ranging sensor, the topography of the exposed hemisphere is defined. With this information, the intelligent robot system delineates several picking zones on the tree, and an appropriate approach vector perpendicular to the target plane for each zone is defined (refer to Figure 2).

The robot now prepares to harvest oranges in each of the zones. The arm is moved to the appropriate zone approach point and a local picture is taken to identify and locate oranges in the specific picking zone. Identification includes finding all oranges, even those which are partially hidden by leaves. The location task includes defining the x, y location of the center of each orange identified. The distance to each orange is determined using a single point range sensor. For each orange the robot arm then moves to the appropriate x, y location and approaches the orange for acquisition by monitoring range. The arm avoids branches and other objects by using an array of proximity sensors. Upon reaching the orange, the robots multi-fingered hand grasps the orange and pulls it off the branch. Damage to the orange is avoided by sensing the grasping and picking forces. The orange is then dropped into a sleeve or bag for collection. All the oranges in the harvesting zone are picked in this fashion. The arm then moves to the other zones until the entire hemisphere has been picked.

Intelligent Robot System Requirements

There are many implications in the procedure just described. Various types of sensing will be required. This includes both the sensing hardware and the algorithms to interpret and make decisions based on the sensory data. Non-contact sensing includes vision, range, and proximity sensing, while contact sensing includes force and tactile sensing. The non-contact sensing requirements will be:

Vision - Required for global, 2-D definition of the tree perimeter and orange identification and location.

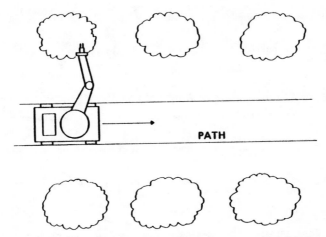

Fig. 1   Intelligent Robot Orange Picker System

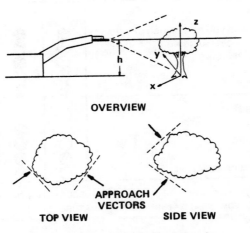

Fig. 2   Zone Definition and Mapping

Range – Dense ranging required to define hemisphere topography and zone approach vectors, and single point ranging required for orange acquisition control.

Proximity – Required for collision avoidance.

The contact sensing requirements include:

Force – Required for detecting when orange releases from branch.

Tactile – Required for monitoring grasp force when acquiring each orange and detecting slip.

In addition to the sensing requirements there are many system requirements such as:

Mechanincal Functions – The arm itself must meet the requirements for size, weight, flexibility, and positioning accuracy.

Mobility – The system must navigate from one set of trees to the next.

System Integration – The sensing and arm control functions must be tied together into a closed-loop system.

Table 2 shows the relative state of maturity for the robotic orange harvester sensing requirements today, in 1985 and in 1990.

Table 2.   Summary of Relative Maturity of Robot Orange Picker Sensing Requirements

	1983	1985	1990
SENSORS			
HIGH AND LOW RESOLUTION CAMERAS	●	●	●
SINGLE-POINT RANGING	◍	◉	●
DENSE RANGING	○	◉	●
TACTILE	○	◍	●
FORCE	◍	●	●
ALGORITHMS			
VISION	◍	◉	●
RANGING	○	◉	●
FORCE	◍	◉	●
SYSTEM INTEGRATION	○	◍	●

○ VERY IMMATURE   ◉ ADEQUATE FOR THE JOB
◍ SOMEWHAT MATURE   ● MATURE

In general, vision sensors (cameras) available today are comparatively mature with various resolution cameras being commercially available. However, developing algorithms to deal with orange identification and location represents a difficult problem since the system must distinguish oranges which may be partially hidden by leaves. It may be that infrared or color sensing will be required to discern oranges from the background. This orange identification and location problem has some similarities to the kinds of problems being addressed by researchers looking at unstructured factory assembly tasks. Numerous groups are currently working this kind of vision algorithm development. In an unstructured assembly environment the robot will be required to view randomly placed parts in a bin or tray and to identify a specific part of interest. At least in the case of oranges, the data base is relatively small, since the goal is to distinguish oranges from everything else. At any rate, vision algorithm development beyond today's capabilities will be required for reasonable operation of the robot orange harvester.

Two types of range sensing will be required: single point and dense (multiple point) ranging. Numerous techniques have been proposed for single point range finding, including lasers, acoustic sensors and ultrasonic sensors. These range finders have been shown to be accurate to within approximately 1 mm. More recently Honeywell has developed a solid state device which has been shown to provide a resolution of better than 0.5 mm at close ranges. There are a few techniques under development for providing dense range maps, however, dense range mapping is comparatively immature. Also, there have been few robotic applications using range sensing to date, so that algorithm development for ranging needs much work.

Contact sensing requirements include force and tactile sensing. Wrist force sensors can detect the magnitude and direction of the force occurring on the arm. There are at least three commercial wrist force sensors available today which would be adequate for the robot orange harvester. Tactile or fingertip sensing is comparatively immature, although there are a few groups working in this area. The concepts are primarily pressure sensitive grids which can be mounted on robot gripper fingers. Much work still needs to be done in this area. Force system algorithm development is somewhat mature. A couple of different groups have looked at tasks such as part mating using force sensing feedback. Little or no work on algorithm development for tactile sensors has been performed, although it is felt that some of the current vision processing techniques may be applicable.

A major system integration effort will be required to tie all the sensing functions together into a closed loop system. There are various groups investigating closed loop robotic sensor systems. However, integration to the level required for the robotic orange harvester has not been demonstrated by anyone to date. Honeywell is currently working on a program to integrate vision, range, force and touch sensing in a robotic assembly system. Demonstration of this system is scheduled for the 1986 time frame and will represent a significant step forward in system integration technology.

The remaining system considerations include the mechanical functions and mobility. The mechanical functions required for this task do not represent a technological hurdle. Many industrial robots are currently available with six or more degrees of freedom. The best available positioning resolutions for today's assembly robots are around 0.05 mm. The orange harvesting task should not require this degree of accuracy.

Different degrees of mobility for the robot orange harvester could be obtained. Initially, the system could move on tracks throughout the grove. This would be relatively straightforward to implement. Eventually, the system could become completely autonomous. We are a long way from reaching this kind of capability, although there are a few groups working in this area.

Performance/Cost Comparisons

In order to get a quantitative feel for the feasibility of the postulated robot orange picker, some rough assumptions and estimates were made about potential system costs and performance capabilities in three time frames: today, 1985 and 1990.

Figure 3 shows system cost estimates for the intelligent robot orange picker. Today, it appears that the state of technology would allow possibly a one-arm development system. This system could cost well over a half million dollars. By 1985 a two-arm system may be possible with maturing technology at a $200,000 cost. By 1990, the required technologies should be quite mature and a cost of $50,000 (in 1983 dollars) may be obtainable on some sort of quantity build level.

Figure 4 shows performance estimates for orange picking rates per hour. The 1600 oranges per hour rate for a human picker is shown for reference. With the 1983 one-arm robot development system, the performance would undoubtedly be marginal. A good estimate for picking rate might be two oranges per minute or 120 oranges per hour. The 1985 system should do better with two arms and smarter sensing algorithms. A doubling in picking rate to 240 oranges per hour seems reasonable. By 1990, with increased intelligence and maturing technologies a large increase in performance is predicted. It appears that picking rates of one orange every two seconds could be obtained. This translates into 1800 oranges per hour, exceeding the human picking rate.

Based on these estimates a payback analysis for 1990 was performed. With a state-wide six month harvest season, a full time human orange picker working a 40 hour week harvests 1.6 million oranges for the season and should earn something over $11,000 for the season. Assuming the robot works a 24 hour day and a five day week, the robot will harvest 5.6 million oranges for the season. This is 3½ times the total production of the human picker. In 1990, recall that a $50,000 capital investment (1983 dollars) for the intelligent robot orange picker is required. It is assumed that some supervision of the robot is required. In today's factories that use robots, generally one supervisor is required for every four robots, and this is assumed for the orange picker. Also, it is estimated that the supervisor will be paid $15 per hour. Therefore, seasonal supervision of four robots working 24 hour days comes to $46,800 or $11,700 for each robot. Also, with $1000 estimated for maintenance each year, the annual robot system cost is $12,700. Since the robot replaces 3½ human pickers, the annual labor savings is 3½ times $11,000 or $38,500. Therefore, the annual net savings is $25,800, implying a two year payback for the $50,000 capital investment.

Conclusions

It is understood that this analysis relied heavily on a number of far reaching assumptions and predictions of the future. The value of this exercise lies in the structuring and analysis that it requires. Orange harvesting was selected initially because it appeared to defy automation. However, whether it be oranges or apples or lettuce or tobacco or any other major sector of agriculture, such an exercise has merit. There is value in defining the labor-intensive tasks that are being performed today, for these are the candidate tasks for the intelligent robot systems that will be maturing by the end of the decade. An analysis such as this one will surely yield a few surprises and also some new insights.

REFERENCES

1. Harmon, Leon D., 1982, Automated Tactile Sensing, The International Journal of Robotics Research, Vol. 1, No. 2, Summer 1982; 3-32.

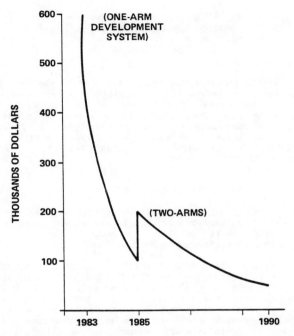

Fig. 3   Intelligent Robot Orange Picker Cost Estimates (1983 Dollars)

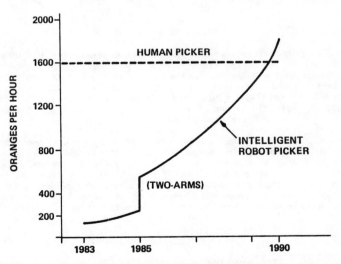

Fig. 4   Intelligent Robot Orange Picker Performance Estimates

111

ROBOTIC HARVESTING OF APPLES

A. Grand d'Esnon

A machine, with vision guidance, is being developed at the French Institute
of Agricultural Engineering Research, Montpellier France, to harvest apples
with minimal manual assistance. The machine was designed to locate an individual fruit in a two dimensional plane, direct a telescopic arm along this
planar point until contact with the fruit is made, pick the fruit and automatically deposit the apple in a conventional harvesting bin. A prototype
machine is scheduled for completion and initial laboratory testing during
the latter part of 1983.

SYSTEM DESCRIPTION

A diagram of the robotic harvester is shown in Fig. 1. The harvester consists of three major components; a telescopic arm with a suitable end effec-
tor, a charge coupled device (CCD) camera mounted along the axis of the
telescopic arm, and a microcomputer control system.

The telescopic arm (which is a hollow tube) is held by a barrel that follows
a movement of translation to the fruit. The barrel is used as a gliding
channel that is slewable in a plane tilted slightly above the horizontal.
It can reach any point in the same level of the tree's canopy. The barrel
is held by a carriage sliding in the vertical direction. Attached to the
carriage is the CCD camera which can scan the entire tree searching for
harvestable fruit. The microcomputer control system coordinates the scans
of the camera, fruit detection, and activation and control of the telescopic
arm.

THEORY OF OPERATION

While the carriage moves up along the vertical framework the camera scans
successive horizontal strips (1000 x 10 mm) producing a search pattern in a
vertical plane moving from the bottom of the canopy to the top. To improve
the reliability of the fruit detection operation filters in two wave
lengths, 960 mm and 740 mm are used. When more than five adjacent fruit
points are detected the computer initiates movement of the arm such that it
is directed to the center of the planar fruit location.

A sensor on the end of the arm detects when the arm has reached a fruit.
The computer then stops arm movement and initiates the picking operation.
The apple is bitten with a flexible cup using a flap or an articulated plastic hand. As the arm retracts, the fruit is detached from the branch and
rolls down the hollow arm through a deceleration device and into the bin.
If the apple cannot be harvested, because it is out of the work space of the
device or obstructed by a branch, the arm is automatically retracted.

The author is: Research Engineer, CEMAGREF Av du val de Montferrand,
Domaine de La Valette, BE 5095 34033 Montpellier Cedex 67.524343.

APPLE HARVESTER

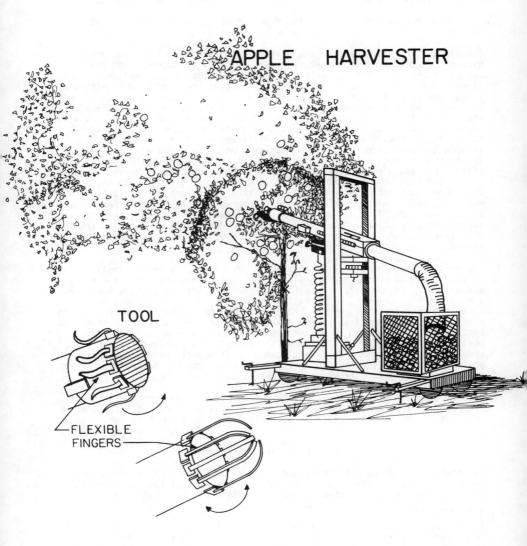

TOOL

FLEXIBLE
FINGERS

Fig. 1 Robotic Apple Harvester

113

CONTROLLING AGRICULTURAL MACHINERY INTELLIGENTLY

Clarence E. Johnson, Robert L. Schafer, and Steven C. Young [1]/

Member ASAE   Senior Member ASAE   Assoc. Member ASAE

The Agricultural scene within the United States and around the world has
many common characteristics and yet there are many variations from one
region to the next, particularly in terms of climate, soils, and crops
grown. And we are becoming increasingly aware of the need to deal with the
variations that exist on a more local level. If we desire to optimize the
efficiency and effectiveness of our machinery systems, then we must deal
with the variations that exist in agricultural crop production, which places
additional requirements on the functional performance and design of agricultural machines. Frequently we see different machinery systems used to
produce the same crop in different regions, and this trend will increase.
In addition to regional variations there are often variations within a
field. For example, Warrick and Nielsen (1980) report that the coefficient
of variation of some soil physical properties may be greater than 1000%
within one soil type within one field. Large variations within a field may
require that the machine be "adjusted" on-the-go so that optimal performance
is achieved. Machines and operations that require frequent adjustments to
satisfy demands of performance are prime candidates for automatic control.
Many functions of future agricultural machinery will be subject to automatic
control.

The need for automatic controls in machinery systems has been recognized for
a long time, and automatic controls in one form or another are not new to
agriculture. Mechanical, hydraulic and electro-hydraulic systems have been
around for some time. The increasing interest in controls in this era
stems from the invention of the transistor and the resultant developments in
electronics and computers. These developments provide us real opportunities
that we could only dream about just 10 years ago. We now see electronic
moni toring and control equipment which incorporates computers to perform
complex assessments and analyses. Lanphier (1982) discussed some specific
applications of electronic and electro-hydraulic control systems in current
and future agri cultural production systems.

In this paper we will first summarize our involvement and work on automatic
controls and then present some of our perceptions and concepts for automatic
controls in the future, including possible impact on production agriculture.

A GLIMPSE INTO THE PAST

Researchers at Auburn University and the USDA National Tillage Machinery
Laboratory became interested in automatic control--in particular, automatic
guidance- when they initiated Controlled Traffic tillage studies on highly

1/ Professor, Agricultural Engineering Dept., Alabama Agricultural
Experiment Station, Director, National Tillage Machinery Laboratory
USDA-ARS, and Graduate Research Associate, Agricultural Engineering Dept.,
Alabama Agricultural Experiment Station, Auburn, AL, respectively.

114

compactable Coastal Plains sandy and sandy loam soils (Dumas et al., 1973). The basic idea of Controlled Traffic was to create and utilize definite traffic zones and crop production zones. An effort was made to keep the two zones as separate as possible because the optimum conditions for traction, transport, and mobility and those for crop production are in conflict. The ability of the operator to determine the past traffic zone and to stay in that zone was soon realized as a limiting factor. An operator would wander back and forth over the desired path by at least +15cm at field operating speeds. Thus, the traffic zone was increased and the crop production zone was decreased, which was quite undesirable.

Automatic Guidance

Schafer and Young (1979) developed an automatically steered tractor that was guided by a buried wire circuit to aid the controlled traffic research. Their controller was a discrete logic "hardwared" device with no computer intelligence; they utilized the contemporary electronics of the time.

The results of the Controlled Traffic research showed the effect of field traffic on soil compaction and the detrimental effect of soil compaction on crop productivity. These results coupled with the expansion of computer technology added impetus to automatic control and automatic guidance research. Young et al. (1983) developed a computer-based guidance system for a tractor that replaced the discrete logic controller developed by Schafer and Young (1979). The computer-based system sensed the positional error from a buried wire and decided how to guide the tractor. This guidance system incorporated a speed sensor to effectively change the steering rate based on speed.

Research results showed that maximum error from a straight line path was within +3 cm at field speeds; an acceptable error. One of the interesting results of this research was that only about 5% of the microcomputer's time was required to guide the tractor; thus, the other 95% of the time could be employed to enhance production, increase energy efficiency, and conserve resources in field operations.

Smith (1982) realized that a buried wire path was not desirable because of cost and lack of versatility. He developed the algorithms necessary for guiding a tractor through the field if, instead of a buried wire, a generalized spatial position-sensing system was available. Algorithms were developed for three tractor-implement configurations--a front steer tractor-trailed implement, an articulated steer tractor with a trailed implement, and a rear steered machine with a front mounted implement (eg. a self-propelled combine or cotton picker). These algorithms require that the position of two points on the tractor be measured by a precise spatial position-sensing system.

Smith (1982) verified the algorithms with a computer simulation and with a physical model in our Vehicle Guidance Laboratory. This laboratory includes a 1/9 scale model tractor under computer control for studying automatic guidance concepts (Young, Schafer, and Johnson, 1979).

Verification of the algorithms in the field has not been accomplished because an adequate spatial position-sensing system has not been available.

As we developed experience in the use of computers for automatic guidance and realized the computational power that was available to us for control applications, we became interested in future aspects of controlling agricultural machinery intelligently.

A VIEW INTO THE FUTURE

Land resources for crop production are limited. Thus, most future increases in crop production will probably be based on top management techniques in which all aspects of crop production are integrated. So the most desirable environment for plant growth within environmental resource constraints must be developed and maintained. Engineering researchers, developers and designers face the crucial question: Is it possible to develop equipment that can be used everywhere that will provide optimum results everywhere? This question is compounded with variations in soils and crops in the field that can be denoted by various electromagnetic or photographic sensors in satellites. Thus, our hypothesis is: Future machinery used in production agriculture will be automatically controlled to prescribe cultural practices, based on soil, crop and climate. Some soil and crop information may be sensed on-the-go and stored in a computer on board the prime mover or field machine. This computer, in turn, could be programmed to make real time decisions based on this information to control cultural practices such as fertilizer, herbicide, and pesticide application. Tillage and planting operations could be controlled also.

We have developed a tillage concept called Custom Prescribed Tillage or CPT (Schafer and Johnson, 1982; Pratt, 1983). In this concept we envision that tillage will be prescribed (see Fig. 1) based on the dynamic specifications of crop needs and our knowledge of technology to meet those needs and resources and machines available will then be drawn upon as needed for the job. To fully develop the concept of CPT we need a body of knowledge of soil dynamics and soil-machine relationships and we need to learn how to effectively communicate with plants--that's right, we must learn how to talk to plants. And, the CPT concept can only be fully implemented through the use of automatic controls.

In the past, we have assumed that uniformity is the rule; in the future, we must assume that variability will be the rule and uniformity will be the exception. Thus, whereas past field operations were primarily a blanket treatment, our future field operations will take variability into account. These variations must be accommodated for as we travel across the field, based on knowledge of what conditions lie ahead in the field. This can be possible by use of a computer control system on board the machinery system similar to that described by Schafer, et al. (1981) and shown in Fig. 2. Important to this concept is a general spatial position-sensing system that can pinpoint the position of a machine in the field at any time. Anticipatory information may include data from digitized maps of soil and crop variation within a field or data generated from signals from trans- ducers on board the machine during a prior field operation or crop season. Anticipatory information may be gleaned from Eros data maps which with current technology may be digitized to an area resolution of less than 1/2 hectare.

In any control system, monitoring is the first step. We must develop the capability to measure crop and soil conditions dynamically on-the-go as we operate equipment in the field. The future automatically controlled machine may receive information from satellites and sense prior soil or crop con- ditions and current or final conditions. Advances have been made in sensor technology that permit critical information to be collected and used in a feedback control system to control cultural operations.

For example, some automatic control of combine functions have been imple- mented by Kruse, Krutz and Higgins (1983). Lee et al. (1983) have de- veloped computer controlled automatic vibrating tillage equipment which attains minimum draft and power. The technology currently under development for continuous and on-the-go sensing of soil moisture and soil physical con- dition now, would make it feasible to control placement of seeds into moist soil and control other operations which are dependent on soil moisture and soil physical condition. However, we need much more sensor technology.

Another deficit in our current technology is our lack of commercially available, reliable, and precise spatial position-sensing systems. Once these systems are developed, site specific information can be sensed and logged into a three-dimensional computer map. The accuracy required from a spatial position- sensing system for developing adequate computer maps may not be as critical as the accuracy required for automatic guidance. Future automatic guidance systems may involve multiple-stage spatial position-sensing systems. However, lack of spatial position-sensing systems in the field should not deter our efforts to pursue continued development of the sensor technology needed to utilize the potential of a spatial position-sensing system.

With adequate sensor technology and spatial position-sensing technology fully developed, a very futuristic scenario is feasible. Existing computer software will permit automatic guidance of the machine so that it follows the same tracks repeatedly without operator intervention. The operator's role will change. He will be observing conditions within the field and the crop and logging this information into the computer data base. For example, he may key the computer to record the location of specific weed, insect, disease or other infestations as the machine passes through these areas. Automatic sensors may sense soil water content spatially and with depth, canopy temperature, soil surface cover, crop yield, and other variables on-the-go. Constant or manually obtained data, such as soil texture, organic matter, topsoil depth, and topography, could be logged into a computer data map as longer term or permanent data.

Automatic controls which intelligently utilize such information would make it possible to optimize the treatment and production of each hectare or the smallest feasible field area rather than broadcast treating the entire field. A yield map would make it possible to fertilize and plant a crop in each square meter of the field according to its yield potential and to fully exploit its yield potential. A weed map and insect map would make it possible to spot treat with a prescribed mixture of specific herbicides or insecticides rather than broadcast treatment of a single mixture. That is, the mixture of pesticides would be dynamically varied to exactly match the infestation. Thus, it could be possible to eradicate weeds and other pests that now get out of control.

Out-of-the field and off-line comparisons between different data sets for the same field may make it possible to diagnose problems before they become disastrous. For example, comparison of a leaf canopy temperature computer map with a computer map of soil moisture may make it possible to isolate areas where plant roots may be having problems. If soil moisture is adequate but leaf temperature is too high, then the plants may be stressed due to a rooting problem. Similar analyses are almost unlimited; in fact, we believe our only limits are our imagination, ability and desire to develop the needed technology.

THE CHALLENGE OF THE FUTURE

The future of automatic controls on agricultural machinery will be governed by factors of feasibility, economics, and reliability. Engineers and scientists must determine and develop the technology that is technically feasible and that will have the greatest impact at any point in time. Development and design engineers must develop the systems that are the most reliable and economically feasible because the consumer--the farmer--will largely base his decisions to utilize automatically controlled machinery on reliability

and economics. The farmer is always interested in improving machinery efficiency and effectiveness, but he constantly lives with the realism of his
pocketbook.

The challenges to the engineering and scientific communities are considerable, the opportunities are unlimited, and the future is exciting. Who
can predict what agricultural machinery systems will look like 20 years from
now?

REFERENCES

1. Dumas, W.T., A.C. Trouse, Jr., L.A. Smith, F.A. Kummer, and W.R. Gill.
   1973. Development and evaluation of tillage and other cultural practices in a controlled traffic system for cotton production in the
   Southern Coastal Plain. Trans. of the ASAE 16(5):872-875,880.

2. Kruse, J., G.W. Krutz, and L.F. Higgins. 1983. Computer controls for
   the combine. Agricultural Engineering 64(2):7-9.

3. Lanphier, R.C. 1982. Looking ahead for the farm equipment industry.
   Agricultural Engineering 60(10):14-19.

4. Lee, K., O. Kitani, T. Okamoto, K. Miura and K. Morimoto. 1983.
   Automatic control for vibratory tillage III: Computer controlled
   vibratory tillage. Jour. of Soc. of Agri. Machinery 44(4):605-610.

5. Pratt, M. 1983. Soil Tillage: The challenge of diversity!
   Agricultural Engineering 64(8):6-9.

6. Schafer, R.L. and C.E. Johnson. 1982. Changing soil condition--the
   soil dynamics of tillage. Predicting Tillage Effects on Soil Physical
   Properties and Processes, ASA Special Publication No. 44, pp. 13-28.

7. Schafer, R.L. and R.E. Young. 1979. An automatic guidance system for
   tractors. Trans. of the ASAE 22(1):46-49,56.

8. Schafer, R.L., C.E. Johnson, S.C. Young, and J.G. Hendrick. 1981.
   Control concepts for tillage systems. ASAE paper No. 81-1601.

9. Smith, L.A. 1982. The development of algorithms for automatic
   guidance of agricultural machines. Unpublished Ph.D. thesis. Auburn
   University, Auburn, AL.

10. Warrick, A.W. and D.R. Nielsen. 1980. "Spatial variability of soil
    physical properties in the field". In: D. Hillel (ed.) Applications
    of Soil Physics. Academic Press, New York. pp 319-344.

11. Young, S.C., C.E. Johnson, and R.L. Schafer. 1983. A vehicle guidance
    controller. Trans. of the ASAE (In press).

12. Young, S.C., R.L. Schafer, and C.E. Johnson. 1979. A microcomputer
    based vehicle guidance simulator. ASAE Paper No. 79-1619.

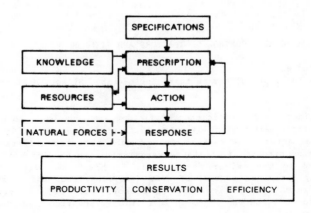

Fig. 1 Custom Prescribed Tillage (CPT).
(NTML Photo No. P10,332a)

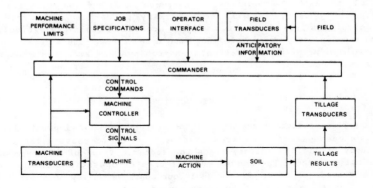

Fig. 2 A control concept for tillage systems.
(NTML Photo No. P10,332b)

AUTOMATIC CONTROL OF TRACTORS AND FIELD MACHINES

M E Moncaster   G O Harries
National Institute of Agricultural Engineering, Silsoe, Bedford, UK

The need to increase the efficiency of field operations leads to a need for improved control. Although fully automatic control is technically feasible, it is not yet acceptable to farmers and manufacturers and attention is therefore being directed initially towards improving the information which is presented to the driver to help him in his task, with a view to incorporating automatic control of some functions later before attempting full automation. A wide variety of sensors will be required, and a microprocessor will be needed to collect information from them and process it for display and control purposes. Investigations with a simulated system suggest the need for a digital serial data communications system using a common communications bus. Based on these investigations a prototype instrumentation system has been installed on a working tractor for carrying out field trials. If comprehensive instrumentation and control systems are to be widely used, it will be necessary to derive a standard data communications interface which will be accepted within the agricultural engineering industry and development should begin now. There is also a need for development of interfaces between such systems and farm management computers.

INTRODUCTION

For many years a significant part of the research and development effort at NIAE has been directed towards finding ways of increasing the working efficiency of field machinery since the economic value of improvements is potentially large. The annual cost of tractor operations in the United Kingdom exceeds £1,000 million with labour costs amounting to about a third of that figure. Similarly, the operation of combine harvesters costs about £200 million, although the labour cost is only about 5% of that figure. The cost of operating other types of self-propelled machines is relatively small since their numbers are not large. It follows that even modest improvements could lead to very substantial national savings; increasing the efficiency of tractor operations by only 1% could save more than £10 million annually. In addition, improvements in work rate allow better timeliness which may well represent even greater saving although it is difficult to make reliable estimates.

One approach which has been followed in recent years was to explore the possibility of complete automatic control of a tractor, including driverless operation. On the assumption that some supervision would be necessary, it is potentially feasible for one man to achieve say two or three times the previous work rate and to save 15 or 20% of the total operating cost.

Operational Research studies suggested that the extra capital cost in the automatic control system would be acceptable provided that it did not exceed about 15% of the capital cost of the tractor.

For the purposes of a practical study it was decided to concentrate on the automation of ploughing and a system was developed and tested in the field. The tractor was fitted with a fully automatic system for raising, lowering and inverting a reversible plough, for turning at the headland using an optical ranging system, and for following an existing furrow. Thus, one pass was required with manual control but thereafter the tractor would plough the field automatically. Development was pursued to the point where the system performed satisfactorily for more than 95% of the time; in the remaining cases it failed to detect the previous furrow on completing its turn at the headland and was stopped automatically by a safety system. Unfortunately, although the system attracted wide interest, this did not extend to any commitment from the tractor manufacturers and further development of the system stopped, about two years ago.

Analysis of possible reasons for the lack of commercial interest suggested that the steadily increasing average power of tractors (40 kW to nearly 60 kW in the UK over the past 20 years) had led to increases in work rate and hence reduced the demand for automatic operation. However, it seems likely that the main reason is that the market was not, and probably is still not, prepared to make the large step to fully automatic working. Consequently, it was decided to proceed in a series of smaller but linked steps, concentrating initially on finding ways of providing the driver with better information so that he could improve the efficiency of his work, then providing automation of selected functions but with a manual over-ride, before considering full automation again.

This paper describes current work on the first of these steps, using a systems approach to the instrumentation and control of field machinery, which is being directed in the first instance towards tractors and implements. Since there are 400,000 tractors in the UK compared with 40,000 combine harvesters and a smaller number of other self-propelled field machines it is reasonable to begin by looking first at tractors. However, other projects at NIAE include the development of control systems specifically for field sprayers and for a purpose-built farm transport vehicle, although these will not be described here.

SYSTEMS APPROACH

The purpose of providing instrumentation and control is to increase the effectiveness with which field operations can be carried out. In the case of a tractor and implement it would be possible to consider the tractor and implement separately and attempt to optimise their performance independently. However, since they interact with one another, very strongly for example when a tractor is carrying out a heavy draught operation with the implement coupled to the tractor via a three-point linkage, this approach will not lead to optimum performance of the tractor-implement combination. It is therefore essential to treat the tractor-implement combination as a system. Further, if manual control by the tractor driver plays a significant part in the operation being carried out, the driver must be considered as part of the system which is to be optimised and careful attention to the ergonomics aspects will be required.

GENERAL REQUIREMENTS

The main elements in the instrumentation and control system, are

- sensors for collecting relevant data from the tractor, the implement and their environment;

- a microprocessor or microprocessors for processing data to derive and display information to the driver and to derive control signals;

- actuators for putting control signals into effect;

- a means of displaying information to the driver in alpha-numeric and graphical form;

- a keyboard or other means with which the driver can input information, select displays etc;

- a communications system to carry data from sensors and other input devices to the microprocessor, and from the microprocessor to actuators;

Especially on the smaller farms (less than 100 hectares) which may have only three or four tractors, a tractor is likely to be used for a variety of different tasks. Surveys have shown that for a 50-60 kW tractor, the mean number of implements used in a year is 18, and that a different implement is hitched to the tractor on average 94 times or nearly twice a week on average. Thus, if the instrumentation and control system is to achieve optimisation of the performance of the tractor-implement combination, it must take into account at least the needs of all the commonly used implements and the most common operations. This implies the need for considerable versatility, since the alternative of providing separate systems for each type of implement and operation is clearly impracticable.

Needless to say the system must be capable of working in the relative hostile environment, including the electromagnetic environment, found on a tractor, and the size and shape of the components must allow convenient installation. The latter is particularly important in the case of components which must be mounted within the cab.

Displays and keyboards must be mounted so that they can be conveniently used by the driver, with minimum movement; displays must be clearly visible under all likely ambient light conditions.

PARAMETERS TO BE MEASURED AND DISPLAYED

Table 1 shows examples of the main parameters which require measurement and display, indicating those which must be derived from two or more input parameters. It also suggests those functions which require control and may be appropriate for automation at a later stage.

Table 1.
==

Machine	Parameter to be measured	Parameter to be displayed		Control functions
Tractor	Engine speed	Slip	(D)	Transmission
	PTO torque	PTO power	(D)	Engine
	Fuel consumption	Specific fuel consumption	(D)	management
	Forward speed	Forward speed		
	Draught	Draught power	(D)	
	Temperatures	Total power	(D)	
	Oil pressures	Area worked	(D)	
		Excess temps		
		Pressure out of limits		
Plough	Depth	Depth		Depth
	Draught			Setting
Cultivator/ Harrow	Depth	Depth		Depth
	Draught			Geometry
Rotary Tiller	Torque	Depth		Rotor
	Depth			speed
Drill	Depth	Depth		Depth
	Soil moisture	Moisture		
	Seed hopper content	Reserve	(D)	
	Seeding rate	Blockage		
	Speed	Seed, total area	(D)	
Sprayer	Speed	Application rate	(D)	Flow
	Flow	Area covered	(D)	
		Reserve	(D)	
Forage harvester	Throughput	Blockage warning		Forward speed
Baler	Throughput	Blockage warning		Forward
	Bale length	Work rate	(D)	speed
	Bale density			
Fertiliser Spreader	Forward speed	Application rate	(D)	Forward
	Output	Work rate	(D)	speed
	Hopper content	Reserve	(D)	Spinner
	Distribution	Distribution pattern		speed

==
Note: D = Inferred value

Many of the sensors required already exist, for example oil pressure or ground speed, and some can be easily adapted from those used in similar applications elsewhere. Some, such as a soil moisture sensor do not at present exist in an appropriate form and may require substantial development.

THE MICROPROCESSOR

The role of the microprocessor is to:-

- organise collection of data from the sensors and from manual inputs via the keyboard;

- process that data to derive information for display or control purposes;

- control data communications;

An increasing number of tractors on the market today have an on-board microprocessor for monitoring and controlling functions on the tractor itself, such as fuel injection, oil pressure etc, for displaying monitored information to the driver and operating alarms, warning him of low oil pressure, high temperature etc. The instrumentation and control system for a tractor-implement combination could in principle use a single microprocessor for all tractor and implement functions. Ultimately this may well be the case, but in view of the advanced state of development of the tractor function system and of the rather special requirements for the control of some of the functions of the power unit, it is likely that the monitoring and control of functions directly associated with the implement and the task in hand will be handled by a second microprocessor. It is possible that more than two microprocessors will be used especially if it proves advantageous to mount one (or more) on the implement.

DISPLAY AND KEYBOARD

Because space within the cab and on the instrument console is limited it is advantageous if not essential to use only a single display and a single means of inputting data manually into the system, such as a keyboard. If the system is to be fitted to an existing tractor then it will be preferable to replace existing instrumentation.

The display should be capable of presenting graphical as well as alpha-numeric information in order to convey relatively complicated information as easily as possible. For example, the working point on a tractive efficiency versus slip curve can be easily interpreted but to convey the same information in alpha-numeric form would be difficult.

COMMUNICATIONS

Consideration of the wide variety of implements which may be used and the numbers of associated sensors leads to the conclusion that data communications presents a very real problem. Some of the desirable features are evident, for example, the need for adaptability to accommodate at least the expected number and types of sensor or the need to minimise as far as possible any additional tasks required of the operator in using the system. In order to explore such features and to test ways in which they might be implemented, a small bench-top simulator was developed.

The simulator was microprocessor based and equipped with a simple keypad, a liquid crystal display capable of simple graphics as well as alpha-numerics, and simulated inputs from sensors of ground speed, engine speed etc.

The development of this simulator led to the following conclusions:

- Sensors should be polled sequentially according to a predetermined program, the rate of interrogation of individual sensors depending upon their function, since parameters which change relatively slowly require less frequent interrogation than others;

- The keyboard should be polled in the same interrogation sequence since it can be regarded as another data inputting device;

- Data communications should be digital and serial, using a full duplex system and the concept of a common highway or bus. A practical realisation could be a common bus which runs round the tractor and the implement, or it could be a star network or some combination of a bus and a star network, depending upon the ease with which cables can be installed;

- In consequence of the common bus type of communication, each sensor must be uniquely identifiable so that the microprocesor can recognise the source of the data it receives. This should be accomplished by encoding the electrical signals from each sensor;

- The system should be designed such that the microprocessor can recognise automatically which particular implement is hitched and, in default of any manual input via the keyboard, presents the highest priority data appropriate to the operation;

- Fault conditions and appropriate alarms should override both selected and default displays.

FIELD TRIALS SYSTEM

Based on the experience gained, a full scale version of a prototype system has been designed and is currently being installed on a tractor, so that the general principles can be demonstrated and development of the system and its component parts can continue with the help of field trials.

As in the simulator, the system is based on a microprocessor and has a simple keypad and a liquid crystal display for presenting alpha-numeric and graphical information. A common bus data communications system is used and the sensors are self-identifying.

At present the working tractor is only fitted with a plough and the combination has sensors for engine speed, gear ratio, ground speed, implement depth and fuel reserve. However, simulated inputs can be used to monitor other operations such as spraying. For the working tractor-plough combination, the display presents:

- a graphical representation of slip versus tractive efficiency showing the working point;

- alpha-numeric displays of implement depth, ground speed, slip, work rate, fuel reserve   expressed in hours of use at the current rate, and real time.

INTERFACES

Cables will be needed to carry data and control signals and power supplies between the tractor and its implements. Obviously electrical connectors will be needed to allow making and breaking of the cable connections whenever an implement is hitched or unhitched. It is equally obvious that the connectors and their pin connections must be compatible for all tractor-implement combinations which will use the instrumentation and control system. In other words a common physical interface is required.

However, for the type of versatile system which is envisaged and which has features such as digital serial data communications and automatic identification of sensors, it is clear that the definition of the interface requires a complete description of the electrical form of the data and control signals, including their polarity, amplitude, timing, the number of channels etc, in the same way as interfaces between computers and their peripheral equipment are defined.

Farmers almost inevitably use combinations of tractors and implements made by different manufacturers, and will continue to want to do so when sophisticated instrumentation and control systems of the type envisaged are available. If such systems were to be fitted after manufacture of the tractors and implements, compatibility need not present a problem. However, retrofitting must be regarded as an interim measure only since it cannot lead to the most economic and effective systems. Thus, if the full advantage of these systems is to be realised, a standard definition of the interface must be derived and accepted within the agricultural engineering industry.

It may be argued that it is premature to attempt definition of this interface before considerable experience is obtained in the field with a variety of systems. However, the longer the definition is postponed, the more difficult it will become to obtain acceptance within the industry since many manufacturers, and especially those tractor manufacturers who also offer a wide range of implements, will have become committed to their own individual interface.

As indicated previously, at NIAE the likely numbers and types of sensors, the precision required in the measurements of performance and other prameters, and the rates at which measurements must be made are being considered. From these studies, a picture is emerging of the functional requirements of the data communications system. When this functional specification has been crystalised it will be possible to propose a technical specification and thence a proposed definition for the tractor-implement communications interface.

A comprehensive instrumentation and control system will require inputs of some types of information from the driver, for example the rates of application of spray materials or of fertilisers, and will of course output a wide variety of information to him. At least some of the input information will be derived from farm management decisions, and similarly some of the output information will be relevant to the formulation of farm management decisions. It follows therefore that if a farm management computer is used, there is a real advantage in providing facilities for direct transfer of some types of information between the instrumentation and contol system and the farm management computer, and that an interface between them is required.

Such an interface would relieve the driver of the task of transferring information to and from the farm management computer, and would also provide a simple means of conveying messages to him.

Although it would be possible to provide an on-line connection by using radio, it is more likely that information would be exchanged say only at the beginning or end of the working day, perhaps by connecting the tractor to the farm management computer by cable overnight.

CONCLUSIONS

Previous experience suggests that it is wise to proceed slowly towards full automation of farm machinery, concentrating in the first instance on providing the driver with more and better information so that he can improve the efficiency of the operations he is carrying out.

Provision of this information will require careful attention to the ergonomics aspects and to the development of suitable data communications systems. The latter are likely to use digital serial transmission and a common connection bus.

The need for hitching a variety of implements to a tractor and for changing from one implement to another comparatively frequently leads to a need for an interface in the data communications system between the tractor and the implement. If sophisticated control and instrumentation systems are to be widely used, the derivation of a standard interface which is accepted in the agricultural engineering will be required, and its development should begin now.

There is also merit in providing an interface between the control and instrumentation system and the farm management computer.

REFERENCE

Harries, G O, Ambler, B 1981 Automatic ploughing: A tractor guidance system using opto-electronic remote sensing techniques and a microprocessor based controller. J.agric.Engng Res. 26, 33-53

AUTOMATIC COMBINE

G. W. Krutz   M. P. Mailander
Member ASAE   Student Member ASAE

A combine is used to harvest agricultural grain crops. Profitability primarily depends on minimizing harvest time, grain losses and damage, equipment investment and operating expenses. The combine is one of the most operationally complex and costly pieces of agricultural equipment and requires many adjustments which affect the level of performance. Normally, the combine is adjusted to an "average" crop or field condition. Because of the variabilities which exist, harvesting efficiency is often less than optimum. Therefore, continuous, automatic adjustment of combine settings offers the promise of improvements in operating efficiency with correspondingly significant payoffs in overall performance and combine capacity. Because of this potential, the combine is an ideal selection for development of a significantly improved control system.

Recent strides in developing micro-computer electronics adaptable to the harsh environment of field harvesting have made the efforts to improve the control of combines particularly timely. Fortunately, this has occurred concurrently with a rapidly expanding commercial availability of a wide range of electrical, mechanical and hydraulic actuators and sensors necessary for suitable performance of such a control system. The combination of monitoring and analysis capabilities of a micro-computer and the systems which can translate their electronic signals into useable control functions, offers a unique opportunity to improve machine performance.

Human factor studies have shown that any person doing a task over a reasonable time span can accurately control only four input signals. Additional inputs or tasks increase operator fatigue and errors. A combine has many more than four conditions to be monitored.

In addition to benefits to be gained from unburdening the combine operator, there are numerous dynamic machine adjustments that are feasible as a result of the monitoring and control sophistication possible with a micro-processor centered system. For example, by monitoring the loading on the threshing mechanism one can adjust factors such as ground speed and concave clearance to maintain an optimum level of operation. This will result in improved threshing efficiencies and reduced grain damage.

There are significant economic payoffs to be gained by improving the harvesting efficiency of the combining operation. A soybean loss study conducted in 1976 by an AGEN 530 class at Purdue discovered that some farm operations in Indiana were losing up to 560 kg/ha. This amounted to $130 per hectare of lost profit. The same study determined that of this total loss only 130 kg/ha could be designated as due to preharvest conditions.

Optimum machine adjustments and ground speed depend upon many variables including: yield, moisture content of grain and stalks as well as desired harvesting strategy (in the event of imminent bad weather, the operator may

* Authors are Gary W. Krutz, Associate Professor, and Mike Mailander, Research Instructor, Agricultural Engineering, Purdue University.

wish to pursue a strategy which would maximize the acreage harvested even if this required some sacrifice to increased harvesting losses). Because of the dynamic nature of these factors, it is currently impossible for an operator to continually maintain the performance level of which the machine is capable.

Other benefits foreseen as a direct result of the control concept envisioned include: less downtime due to plugging, fewer failures from overstressing or overloading the machine, increased harvesting capacity (resulting from full cut width using steering control), reduced header losses and improved fuel economy. Another area of significant machine improvement would be related to monitoring of factors such as grain moisture and yields. This yield information might be used as a factor in site specific application of fertilizer within a field.

This research project was a joint venture between Purdue University Agricultural Engineering, International Harvester, and TRW, Ross Gear, TEE and Eagle Monitor Divisions. The project consisted of evaluating continuous moisture sensors, mass flow rate sensors and developing a speed control system.

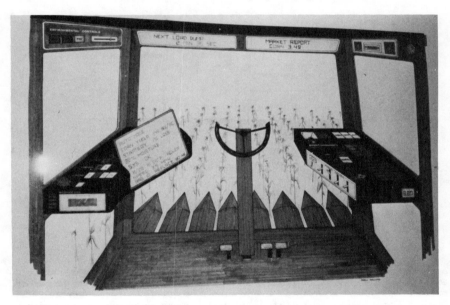

Fig. 1 Proposed Computerized Combine of the 1980's

Ground Speed Control Project One of the project goals was to develop a prototype for a practical combine ground speed controller which utilizes a microprocessor to:

✤ Simplify the operator's routines
✤ Make the combine load-adjustable
✤ Allow the controller to respond to a plugged feeder-conveyor
✤ Maintain optimum travel speed and combine loading

An International Harvester 1460 rotary combine was first equipped with a PDP 11/03 computer system (1). Analog-to-digital (A/D) and digital-to-analog (D/A) converters, digital counters, and latched output control boards provided interfacing with input and output signals. An AED 6200LP floppy disk storage system was connected to the PDP 11/03 to store data and programs.

An Intel Single-Board computer (SBC) 80/05 provided the ground speed control system. The Burr-Brown MP8418-AO board with a bus compatible with the Intel board converted the A/D and D/A signals. The CRT (cathode ray tube) communicated with either the SBC or the PDP, with a switching box to control that choice.

Other devices used to implement the field testing included two momentary contact switches on the microprocessor box to permit its reset. This allowed interruption of the ground speed control program when changing acceleration and deceleration rates. Three potentiometers on the top of the microprocessor box simulated three input voltage signals: commanded ground speed, feeder conveyor pressure, and adjustable target engine speed.

The target engine speed was the center of the speed controller's deadband zone. Two SPDT toggle switches connected the A/D converter board to the potentiometers during testing and calibration or to the actual sensor outputs during field operation. During testing and evaluation all signals except the target engine speed potentiometer value were wired into the PDP; actual ground speed was also recorded by the PDP.

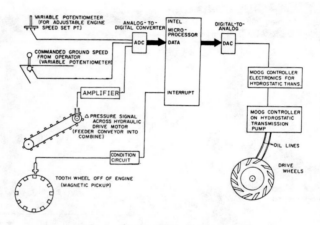

Fig. 2 Controller Input and Output Signals

A signal conditioning circuit was connected to the output of the combine's engine speed sensor to produce a computer interrupt pulse. The number of pulses in 0.33s determined the engine's revolutions per minute. The combine's manually controlled hydrostatic transmission was replaced with a Moog electronic hydrostatic controller. An OEM Controls potentiometer with adjustable level friction replaced the original hydrostatic control lever.

The feeder-conveyor, which originally had been driven by a V-belt pulley from engine power, was modified so that the pulleys could spin freely on two bearings added to the feeder driveshaft. A Ross Gear MAB 08 hydraulic motor was installed to power the feeder driveshaft. A pressure transducer measured the hydraulic pressure needed to turn the feeder.

Field Tests

To begin a field test run using the ground speed system, both the 80/85 SBC and the PDP were started. The PDP executed a program to record sensor data on a floppy disk. With the 80/85, fourteen different control constants were manually entered: acceleration and deceleration rates; deadband zone width;

target engine speed; amount of filtering on feeder pressure and engine speed signals; deceleration rate when feeder becomes plugged; acceleration rate back to full operator command when feeder pressure drops below the "material entering" setpoint; setpoints for the feeder pressure material both for entering and for plugged conditions; the number of times per second the ground speed control loop is executed. Constants could be altered during a field test.

Theory of Operation The control program which evolved maintained engine speed about an adjustable value (target speed) which was set via the variable potentiometer or within a certain range (deadband) about (or around) the targeted harvest speed. When engine speed dropped below the deadband, the control slowed the combine's ground speed. The inverse happened when the engine speed exceeded the deadband.

The operator had absolute control when the combine was not harvesting. As the combine entered the crop and the feeder load pressure exceeded the setpoint for material entering (26 bar), the ground speed control program was activated. The operator's commanded ground speed lever position then determined maximum forward speed. If the operator tried to increase ground speed, the control program offset this and only allowed acceleration at the rate determined in the control program (provided engine speed was above that preset deadband zone). Deceleration occurred if the engine speed was below the deadband zone, but the program could not cause the combine to back up.

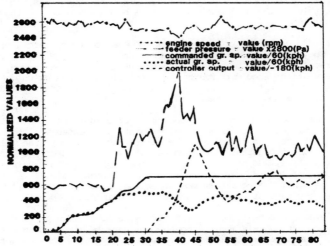

Fig. 3 Wheat Harvest Data, 1980

Engine speed was calculated three times per second. If the average speed was outside the deadband range, the difference between the desired and averaged speed was calculated. An appropriate acceleration or deceleration constant was then calculated and a corresponding adjusted voltage signal outputted to the Moog controller on the hydrostatic transmission drive. If the calculated average engine speed stayed above the deadband, the controls accelerated the combine to the maximum speed set by the operator using his ground speed lever. But if the engine speed remained below the deadband zone, the controls decelerated the combine until it stopped.

When the combine left the crop, the feeder load (pressure) dropped below the 26 bar setpoint to 12 bar, the "no crop entering" pressure. The control system then returned the combine to full operator command speed.

131

The average feeder load also indicated when a plug had stopped the feeder from turning. If feeder pressure exceeded the preset value of 97 bar, controls rapidly decelerated the combine to halt forward travel.

Feeder pressure was used as an on/off switch to allow the operator to move the combine at less than full throttle during unloading, while starting the rotor, or for field movement. The control program provided a smooth automatic transition as the combine entered and left the crop. If the operator had to activate this automatic control manually, it would be an added chore.

To test the plug detection software, the "plugged" pressure variable was set at a low value during corn harvest. By setting the deceleration rate for plugs at a low value of 0.5 kph/s, the combine stopped and started as the crop entered and stopped entering the feeder-proving that the plugging control algorithm was successful.

Because the operator could change the target engine speed, the mass flow rate throughout the combine was controlled. Grain losses exponentially increased with increased mass flow rates. In tests with wheat, grain losses did not determine the maximum flow rate through this combine.

Moisture Sensors The dielectric constant of agricultural materials have been studied for their potential as moisture indicators. The dielectric properties of grain are affected by bulk density, material boundaries, and temperature effects (6). The dielectric properties of many grains are tabulated in the ASAE yearbook.

Moisture transducers would provide useful information to operators or control systems. They will enable wet spots to be sensed thereby allowing compensation in forward speed and harvester adjustments. They may even indicate whether harvesting should continue. One experimental system used as a continuous type sensor with a threshing speed controller in corn (8) and another utilized a batch sensor in soybeans (9).

Most sensing methods are density dependent so care must be taken to avoid differing densities. Since they are also quantity dependent, a constant bulk of the material over the sensor must be maintained. Some of the mechanical methods to insure a constant flow of material over the sensor include passive bleed, active bleed, and plugged flow systems. At Clemson University, a small amount of grain was extracted from the clean grain elevator and fed into a conventional dielectric sensor at fixed time intervals (10). The problem with any type of batch system is the mechanical complexity needed.

Most early dielectric based moisture meters were of a batch type. However the advantages of continuously monitoring the moisture level has long been recognized (11), thus several continuous flow moisture meters are manufactured. Units manufactured by MCS, Inc. were used on the Purdue automatic controlled combine project as well as by Brizgis at the University of Illinois (12).

Four dielectric based continuous moisture sensors were tested in wheat, soybeans and corn during the 1979, 1980, and 1981 harvest seasons. In an effort to control the density of the material passing by the sensors, various configurations were used. Some test results are presented. Each grain moisture sensor experienced unique problems associated with controlling the material flow in the region of the transducer. These sensors showed promise of applicability to the expected use, if the density of the grain is carefully controlled. However their ability to determine corn grain moisture was far superior to their ability to determine wheat or soybean moisture. The sensitivity of these sensors was not satisfactory, especially at lower moisture contents. The sensor intended to measure the moisture content of the material-other-than-grain (MOG) was not a successful indicator.

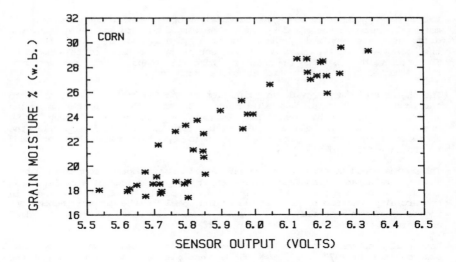

Fig. 4 Moisture Sensor Sensitivity

Mass Flow Sensors The feedrate of material into the combine is one of the independent variables with the greatest effect on combine performance. It is the primary variable used to quantify combine capacity in ASAE standard S396T. The successful measurement of feedrate could provide useful information to help optimize performance, whether to the operator for manual control, or to an automatic controller. In order to develop a useful indicator of feedrate, field tests of several sensors were undertaken.

Previous investigators have attempted to measure the feedrate of material into combines by a variety of methods. Kruse's work at Purdue University utilized the engine speed to indicate the machine load (13). The engine speed is also used in a system that is commercially available for retrofitting to combines (14). A rice combine application in Japan used engine speed to adaptively control the desired straw layer thickness in the feeding portion (15).

Early indications of mass flow are more useful, hence there have been attempts to utilize the header supply auger torque as measured by drive chain force. The attempts in the Netherlands (16), were reportedly successful, although Miles at Clemson (10) had problems with low signal level. As mentioned above, straw layer thickness has been used in Japan (4). The Iseki Model X-HD1500D driverless combine had automatic control of stalk feedrate. Eimer, in Germany, used the displacement of the front feeder hub (18). Mailander used the hydraulic pressure differential on a stationary unit's feeder drive to control feeder speed (19). Friesen, Zoerb, and Bigsby used the threshing cylinder drive torque to indicate feedrate (20).

Besides references in the literature, there have been a number of patents which describe methods of determining material feedrates, primarily in automatic speed control systems. One of the most popular methods is to measure the deflection due to changing thickness of the material mat in the feeder (22). Another patented method utilized the load on the engine through manifold pressure (22). The torque of the threshing mechanism has been measured by various methods (23). One patent describes feedrate determination through a torque measured at the feeder and platform drive (24).

One of the sensors used to attempt to measure the feedrate of material into the combine indicated the torque required to drive the feeder conveyor. This torque was hypothesized to be approximately proportional to the feedrate of material through the feeder area. The torque to turn the feeder conveyor drive shaft was supplied by the output of a hydraulic motor. This replaced the belt and pulley drive on the current production model of this combine. The pressure drop between the motor's intake and exhaust ports was measured by a Sensotec Z/767 differential pressure transducer.

The engine speed may provide an indication of the feedrate of material into the combine. Increasing feedrate into the combine loads the threshing and material handling mechanisms. This loading is transmitted through drive trains back to the engine. The engine decreases in speed to provide the increased torque demand. The resulting engine speed droop can then be used to indicate the feedrate.

The engine speed transducer on the 1460 Axial-Flow combine consisted of a magnetic reluctance transducer which sensed the projections of a pulley turning at the engine speed. This equipment is currently used on combines in commercial production to provide the engine speed for the digital tachometer. Engine speed was measured with a pulse counter by the experimental data acquisition system.

Potential electronic combine feedrate sensors were tested in wheat, soybeans, and corn in the 1979, 1980, and 1981 harvest seasons. A sensor based on the differential hydraulic pressure of the feeder conveyor drive motor successfully indicated feedrate, and test results are presented.

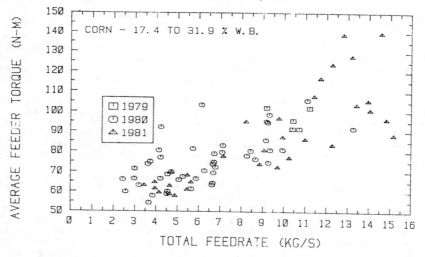

Fig. 5 Feeder Torque vs. Feedrate

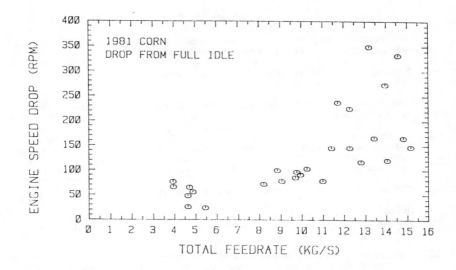

Fig. 6 Engine Speed vs. Feedrate

Engine load, measured in this study by engine speed, provided a very simple indication of material feedrates. Sensors which measured corn header supply auger torque, clean grain auger torque, air pressure in the cleaning area, and the flow of grain into the grain tank were not successful.

Measurements of airflows through the chaffer of an IHC 1460 combine were made during wheat havest in 1982. Air velocity was measured with thermistor-based anemometers at 32 locations on the chaffer. The air velocity through the chaffer provides dynamic information about cleaning system loading such as amount of load and distribution of load (25). Mention of a specific product is for information purposes only and does not imply endorsement of that product.

In Summary

This microprocessor-based control system controlled combine ground speed by utilizing inputs of engine speed and feeder torque, measured by hydraulic oil pressure. The control algorithm used the feeder pressure as an on/off switch to activate the controller when entering or leaving the crop. The amount and type of control was based on the distance and direction the actual engine speed varies from target engine speed. This type of control is proving successful.

Considerable research is needed to improve moisture sensor reliability. The accuracy of dielectric based moisture sensors was dependent upon careful control of material density. Sensor sensitivity was not satisfactory at low moisture ranges.

Machine load due to feedrate was measured by engine speed droop, the torque required to drive the feeder conveyor and the air flow through the chaffer sieve. Feeder torque indicated feedrate of material but was moisture dependent. Engine speed droop was also a good indicator of material feedrate.

REFERENCES

1. Auto—Combine Project (first year report). 1980. Purdue University. Agricultural Engineering Dept. (unpublished).

2. Fardal, R.G. and C.P. Rickerd. 1978. Combine automatic travel control system. US Patent 4, 130,980.

3. Hooper, A.W., and B. Ambler. 1979. A Combine Harvester Discharge Meter. J. Agric. Engng Res. (1979)24,1-10.

4. Kawamura, T., N. Kawamura and K. Namikawa. 1977. Adaptive feed rate control head feeding type combine digital sampled data adaptive control system. Journal of the Society of Agriculture Machinery. Japan. 39(2):157.

5. OEM Controls, Inc., 10 Controls Drive, Shelton, CT 06484.

6. Nelson, S.O. 1973. Microwave Dielectric Properties of Grain and Seed. Transactions of the American Society of Agricultural Engineers. pp. 902-905.

7. Ambler, B., and A.W. Hooper. 1975. An Electronic Recording Instrument For The Grain Flowrate. National Institute of Agricultural Engineering (United Kingdom). Department Note no. DN/C/535/1011.

8. Mailander, M.P., G.W. Krutz, and L.F. Huggins. 1982. Automatic Control of a Combine Threshing Cylinder and Feeder Conveyor. United States Patent No. 4,337,611. July 6, 1982.

9. Baskin, G.R., M.M. Mayeux, and F.E. Sistler. 1981. Meter Design for Onboard Combine Cylinder Speed Readout. American Society of Agricultural Engineers Paper 81-1570.

10. Miles, G. 1982. Seminar Given To The Faculty of Purdue University. October, 1982.

11. Pugh, J.L. 1974. Moisture Measurement Evolution: Sampling to Continuous. American Society of Agricultural Engineers Paper 74-3737.

12. Brizgis, L.J. 1977. Automatic Cylinder Speed Control for Combines. Unpublished Masters Thesis, University of Illinois.

13. Kruse, J., G.W. Krutz, and L.F. Huggins. 1982. Microprocessor Based Combine Ground Speed Controller. ASAE paper no. 82-3040.

14. Ederveen, Jan. 1982. Harvestmor For Harvester Control. Diesel Progress North American. October 1982. pp. 14-15.

15. Herbsthofer, F.J. 1971. Combine Ground Speed and Drum Speed Automatic Control. U.S. 3,609,947. October 5, 1971.

16. Huisman W., J. van Loo, ad J.J.Heijning. 1974. Automatic Feed Rate Control of a Combine Harvester in Wheat. Departmental Report. Department of Agricultural Engineering, Agricultural University, Wageningen, Netherlands.

17. van Loo, Ing. J. 1977. An Automatic Feed Rate Control System For a Combine Harvester. Department of Agricultural Engineering. Agricultural University, Wageningen, Netherlands.

18. Eimer, M. 1974. Automatic Control of Combine Threshing Process. Paper F5 presented at the IFAC Symposium on Automatic Control For Agriculture. Saskatoon, Saskatchewan.

19. Mailander, Michael P. 1979. Computer Control of a Hydraulically Driven Combine Cylinder. Unpublished M.S. Thesis. Purdue University Department of Agricultural Engineering.

20. Friesen, O.H., Gerald C. Zoerb, and Floyd W. Bigsby. 1966. For Combines: Controlling Feed Rates Automatically. Agricultural Engineering. August 1966. pp. 434–435.

21. Andersen, P.B. 1963. Combine Control System. US Patent 3,073,099. January 15, 1963.

22. Pasturczak, S.F. 1953. Self-propelled Harvester with Automatically Controlled Variable-Speed Drive. US Patent 2,639,539. May 26, 1953.

23. Iofinov, S.A., A.V. Tsupak, I.F. Glebushkin, et al. 1977. Grain Harvester Automatic Load Control System. USSR Patent 554832. May 23, 1977.

24. Pool, Stuart D., Edward Svereika, and Tommy A. Middlesworth. 1969. Ground Speed Control. U.S. Patent 3,481,122. December 2, 1969.

25. Streicher, E.A., et al. 1983. Measurement of Combine Cleaning System Airflows. ASAE paper 83-1085.

26. Schueller, J.K., M.P. Mailander, and G.W. Krutz. 1982. Combine Feedrate Sensors. ASAE paper 82-1577.

27. Mailander, M.P., J.K. Schueller, G.W. Krutz. An Evaluation of Four Continuous Moisture Sensors On a Combine. ASAE paper 82-1576.

28. Kawamura, T. 1980. Studies on Adaptive Control of Head Feeding Type Combine. Departmental Report. Kyoto University. (unpublished).

ROBOTIC PRINCIPLES IN THE SELECTIVE
HARVEST OF VALENCIA ORANGES

G. E. Coppock
Member ASAE

The rapid introduction of technological innovations influenced by the Industrial Revolution has until recently resulted mainly in the displacement of
human muscle power from the task of production. The current revolution in
computer and sensory technology may cause an equally momentous change by
applying some of the decision making capabilities of the human brain.

In a brief span of history, mechanization of work has changed the United
States from an agrarian republic into an industrial power. In 1820, more
than 70% of the labor force worked on the farm. Today, only 3% are employed
in agriculture (Ginzberg 1982), yet farmers grow enough to meet the needs of
the entire country and usually with a large surplus to export. In 1850, the
average farm worker supplied food and fiber for 4 people; now each farm
worker provides for 78 people. Much of this enormous increase in productivity can be attributed to mechanization technology broadly defined to include
labor saving devices, fertilizer, pesticides, irrigation, plant improvement,
etc. Its not surprising that mechanization has advanced faster in agriculture than in manufacturing in the United States.

However, all segments of agriculture have not achieved this enormous
increase in productivity. Large numbers of seasonal workers are still
required in the production and harvest of fruit and vegetable crops (Thor
and Mamer 1969). The seasonal aspect of the work in these crops makes it
necessary for workers to migrate to various sections of the country for
employment.

Significant progress has been made in the non-selective harvest of many of
these crops. In non-selective or once-over harvest the total crop is
harvested, the desirable parts are sorted for utilization, and the remainder
is discarded. However, some crops must be selectively harvested. Those
with a young crop on the plant at harvest time or the crop is harvested at
different stages of maturity cannot be harvested in a non-selective manner
without destroying an excessive amount of the product to be harvested at a
later date. Examples of crops in this class are vine ripe tomatoes,
harvested by color; cabbage, harvested by hardness of head and 'Valencia'
oranges, harvested by age of fruit. It is in this area that robotic principles will have application in product detection and removal, whether it is
in harvesting or in grading the product after harvest. Robotic principles
are defined as those principles that take 2 or more inputs and arrive at a
decision.

SELECTIVE HARVEST OF VALENCIA ORANGES

The harvest of Valencia oranges is an excellent example of an operation
where robotic principles may have application. Valencia orange is a very

The author is: G. E. COPPOCK, Professional Engineer, IV, Florida Department
of Citrus, University of Florida, Citrus Research and Education Center,
Lake Alfred, Florida.

important citrus fruit crop in the United States. Economically, it represents about 45% of Florida's orange production with an annual harvest cost of over 100 million dollars.

A unique characteristic of the Valencia orange cultivar is that the fruit reaches maturity suitable for harvest after the young fruit for the next year's crop has formed on the tree (Fig. 1). This characteristic makes non-selective harvesting extremely difficult without causing an excessive reduction in subsequent crop yield (Coppock 1972). Also, to further complicate the problem, young fruit diameter, weight and bonding strength increases over the 3-month harvest season while these properties for mature fruit remain almost constant (Figs. 2 and 3).

Valencia fruit is hand picked by workers who snap individual fruit from the stem and collect it in a bag (Fig. 4). In some areas where the fruit is marketed fresh, it is detached by using a small hand operated clipper. Generally, the worker is in a position to detach (pick) fruit about 75% of the total work time and on the average picks about 40 fruit per minute (Coppock and Jutras 1961).

Selection Criteria

The Valencia orange cultivar has several criteria that can be used for selective harvest. A difference exist between the mature and young fruit in diameter, weight, bonding strength, and color. Several attempts have been made to develop equipment to selectively harvest the mature fruit. The diameter difference was employed in a spindle device (Coppock and Jutras 1960) (Fig. 5). The fruit bonding strength difference was employed in a shaker harvest system where an abscission chemical was used to reduce the mature fruit bonding strength (Coppock et al. 1981) (Fig. 6). These attempts have not resulted in an acceptable method of harvesting Valencia oranges. In this early development work, color was considered as a criteria for selection but was not pursued because of the lack of an effective system of sensing and locating the fruit on the tree.

Methods of sensing the fruit that have been considered are light reflectance, temperature, electrical resistance, capacitance and gamma ray backscatter (Schertz and Brown 1968).

Application of Robotic Principles

Recent developments in robotics makes possible a completely new method for selective harvest (Fig. 7). It consists of combining computer technology with television to identify the fruit and locate the direction coordinates to the fruit. The fruit is identified by comparing the properties of the fruit as viewed by a television camera with properties preprogrammed in the computer. When the properties match, the computer set direction coordinates and directs a mechanical mechanism to move to and retrieve the fruit. By making a change in the computer's program the system can be applied to machines for the selective harvest of other crops. This should be of interest to the high technology industries because of the potential for wide application and its resulting economy.

A potential application of this technology to selectively pick Valencia oranges is shown in Fig. 8. A camera lens for the optical system is mounted at the end of a probe designed to move into the tree to pick fruit. The signals from the scanning camera go to the computer where it is compared with preprogrammed properties. For Valencia oranges, this would probably be color and shape. If a mature fruit is identified, then the computer directs the probe to pick the fruit. Associated equipment would be needed to handle the picked fruit. Because of the relatively slow speed of present computers, several probes operated by one computer would be necessary to approach an economical picking rate.

Several arrangements are possible for positioning the probes in the tree. For operation under present cultural conditions, the operator may have to assist by placing the probe in the general area where the fruit is located on the tree.

Research Needs

The application of this newly acquired technology to harvesting machines will take an integrated research effort by high technology industries, farm machinery manufacturers, commodity industries, and universities. Much of the basic computer technology has already been developed and is being perfected for application in manufacturing, the space program and other related areas. This has been fully documented in other papers presented at this conference. Its application to an agricultural machine, that would be exposed to varying weather conditions and would interface with biological plants with their natural variations, presents many interesting opportunities for exciting and productive research. Several of these research opportunities related to the Valencia harvest will be discussed individually.

Consider the probe design. It must penetrate the tree canopy composed of intermingled limbs and leaves to reach the fruit. Not only must the functional requirements, but the natural biological variation of the tree and the potential for limiting these variations by internal or external means have to be considered in arriving at the optimum design requirements of the probe. Research will be needed to identify the biological variation of the tree shape, location of fruit on the tree and develop methods for limiting these variations. Some of this information is already available but more will be needed.

The fruit detection and probe guidance system present many research opportunities. Although the basic vision technology has been developed, its application to a viable machine has not been demonstrated. Fruit detection depends on making visual contact with each fruit. Methods are needed to assure that visual contact can be obtained under variable operating conditions caused by weather or biological variations. A probe guidance system will be needed that will direct the probe to the fruit at an economically acceptable speed and will provide safeguards against damage to the probe and the tree.

Another area of consideration would be the fruit detachment mechanism. It must detach the fruit in a manner that leaves the fruit in a usable condition. The design requirements of the detachment mechanism will be influenced by the method of collecting the fruit and probe design. Shertz and Brown (1968) investigated various methods of detaching fruit individually.

The fruit collection system design requirements will be influenced by all the other machine components. It will have to be an integral part of the machine. Collecting the fruit from a probe that bring the fruit out of the tree after detachment would be much more complicated than one that detached and dropped the fruit onto a catching surface. A vital consideration in fruit collection is that the fruit is a perishable product and must be salable after it is harvested.

The economics of a machine using robotic principles is an important research area. Assuming that one probe would pick at the rate of one hand laborer while in hand picking position, the hourly rate would be 3000 fruit (a picker is usually in picking position 75% of the time). Also, if a machine accommodated 5 probes, the expected machine rate would be 15,000 fruit/hr. Using one machine operator, the machine would harvest as much as 6.25 hand laborers resulting in a machine/labor ratio of 6.25:1. This ratio compares favorable with a 6:1 ratio estimated for the nearest completing non-selective harvest system (tree shaker) (Coppock 1969). Although the ratios are comparable, the principle advantage of the selective harvester is that there should be no appreciable economic loss from damage to the young crop of fruit.

SUMMARY

Robotic principles have been developed and now are being perfected in manufacturing that opens an exciting new approach to the selective harvest of certain agricultural crops. The possible application of these principles to a machine for harvesting Valencia oranges was discussed and areas needing research and development were explored. To fully develop this technology, it will take the coordinated effort of the high technology industries, agricultural commodity groups, farm equipment manufacturers, and public research agencies.

REFERENCES

1. Coppock, G. E. 1969. Review of citrus mechanization. pp. 777–819. In: B. F. Cargill and G. E. Rossmiller (Technological Implications). Fruit and Vegetable Harvest Mechanization. Rural Manpower Center, Michigan State University, East Lansing, MI RMC Report No. 16.

2. Coppock, G. E. 1972. Properties of young and mature 'Valencia' oranges related to selective harvest by mechanical means. Trans. of the ASAE. 15(2):235–238.

3. Coppock, G. E., H.R. Sumner, and D.B. Churchill. 1981. Shaker methods for selective removal of oranges. Trans. of the ASAE 24(4): 902–904.

4. Coppock, G. E. and P. J. Jutras. 1960. Mechanizing citrus fruit harvesting. Trans. of the ASAE. 3(2):130–132.

5. Coppock, G. E. and P. J. Jutras. 1961. An investigation of the mobile picker's platform approach to partial mechanization of citrus fruit picking. Florida State Hort. Soc. 258–263.

6. Ginzberg, Eli. 1982. The mechanization of work. Scientific American. 247(3):67–75.

7. Schertz, C. E. and G. K. Brown. 1968. Basic considerations in mechanizing citrus harvest. Trans. of the ASAE. 11(2):343–346.

8. Thor, Eric and John W. Mamer. 1969. Rural Manpower-Overview. pp. 51–70. In: B. F. Cargill and G. E. Rossmiller (Technological Implications). Fruit and Vegetable Harvest Mechanization. Rural Manpower Center. Michigan State University, East Lansing, MI RMC Report No. 16.

Fig. 1   Mature and Young Valencia Oranges on the Tree
at Harvest

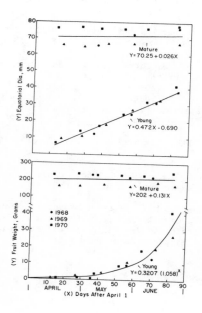

Fig. 2   Effect of Harvest Date on the Weight and Diameter
of Young and Mature Valencia Fruit

142

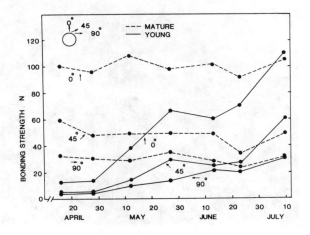

Fig. 3   Effect of Harvest Date on the Bonding Strength of
Young and Mature Valencia Fruit

Fig. 4   Picking Oranges by Conventional Method

Fig. 5 Auger-shaped Spindle Harvesting Device

Fig. 6 Tree Shaker Using the Difference in Weight and
        Bonding Strength Between Mature and Young Fruit
        as Criterion of Selection

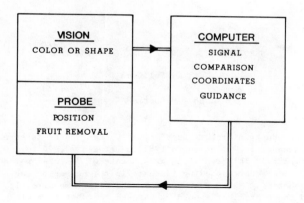

Fig. 7   Schematic of Robotic Sensing
and Guidance System

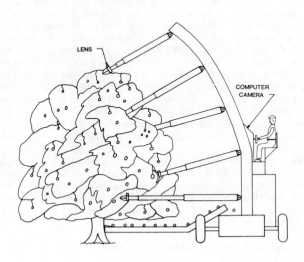

Fig. 8   Schematic of Fruit Picking Machine Using
Robotic Sensing and Guidance System

145

HERO 1 ROBOT: EDUCATIONAL APPLICATIONS

Neil W. Sullivan

Assoc. Member ASAE

The number of industrial robots used in the United States is expected to
increase from under 10,000 units in 1982 to between 75,000 to 150,000 in
1990.(Tanner, 1983) This fantastic growth rate of 33 to 47 percent per
year will require a corresponding increase in the number of engineers
trained in robotics. Many of these engineers will come directly from
colleges and therefore they should be exposed to robot technology,
programing and devices in their undergraduate curriculum. The high cost,
large size and high demand of industrial robots would make these devices
unavailable to most colleges and universities. Therefore an inexpensive
and small scale robot training system would be extremely valuable as a
teaching tool for engineers and other related professions.

THE HERO-1 ROBOT

In January of 1983 Heathkit Education Systems introduced an educational
robot specifically designed as a training system for industries and educational institutions. The Robot called HERO-1 (Heath Educational RObot) was
designed to demonstrate every major principle used in industrial robots but
in a small, inexpensive, and unfrightening unit. The completely assembled
version weights under 18 kg (40 lb) is under 56 cm (22") high and takes up
a floor area of only 56 x 36 cm (22 x 14 in). The robot is energized by
four, 4-amp hour gel cell batteries and comes with a battery charger and
complete operating manuals to explain its features and programing. The
robot can move about independently on its three wheels, has an arm with a
gripper and has a speech synthesizer. In addition, it is equipped with a
number of sensors to detect light, sound, motion and distance. The unit
also comes with a teaching pendant which may be used for rapid programing
of the robot's 8 motors. The robot may be purchased as a kit or fully
assembled and the price, as of September 1983, is $1499 for the kit and
$2495 for the completely assembled version. Heath has also developed a two
volume 1200 page Robotics Education Course which can be purchased for
$99.95. The following sections will describe the individual features of
the robot and how they may be used for educational applications. Figure 1
shows the individual components and features of Hero-1 and will be referred
to in the remaining sections.

Instructor, Department of Agricultural Engineering University of Nebraska,
Lincoln, NE 68583.

This training system can be used to teach the following principles of robotic systems: microprocessor fundamentals and programing, dc motor control and power sources, electronic instrumentation and sensor fundamentals, analog to digital conversion (A/D) and D/A, voice synthesis, and practical limitations and applications for robots. The following sections will describe the features of the HERO-1 and its teaching manuals that can be used to illustrate each of the principles mentioned above.

Microprocessor Fundamentals and Programing

The HERO-1 robot uses a Motorola 6808 Central Processor Unit (CPU). It is very typical of the processors found in commercial applications with some specific characteristics that make it well suited for a robot training system. It is an 8 bit processor capable of addressing 64K (65,536 bytes) of program memory. The CPU board supplied with the robot provides 4k of Random Access Memory (RAM) at addresses 0000 Hex (base 16) to 0FFF Hex. The robot operating system uses the RAM addresses from 0000 Hex to 003D Hex and 0EE0 to 0FFF for jump addresses and timing updates. This leaves RAM addresses from 003F Hex to 0EDF Hex free for user programs. In addition, the unit is supplied with ROM (Read Only Memory) from locations E000 to FFFF Hex. This memory segment contains the robot monitor and initialization routines including some routines that the user program may access such as stored speech phrases and display outputs etc. The remaining memory from address 1000 Hex to DFFF Hex, is not supplied with the robot. An advanced instrumentation class could develop a memory expansion board for the robot as an example project. The procedure necessary to do this expansion is outlined in the technical manual supplied with the robot.

The programing for this robot can only be done manually on the hexadecimal keypad on the top of the robot. (Figure 2) The robot system can use two modes of programing, machine language or robot language. Machine language programing can be used when only the 6808 CPU instruction set is needed for data manipulation and storage. The robot language mode is used whenever external robot devices are being controlled or operated, such as the robot's motors or sensors. This two language mode allows "fast" processing of data and CPU commands in machine language while offering the highly powerful motor control and sensor commands available in the robot language at a significantly slower processing speed. Each of these modes can be entered as desired with simple one instruction commands.

An additional method of programing is provided for using the teaching pendant (Figure 3). This device can be plugged into the robot and is used to manipulate any of the robot's 8 motors manually. These motions may be made either to see the robot's movements and limitations or, by a simple instruction from the keypad, the motions made may be stored for task programing. This is a tremendous time saving feature and is very typical of how industrial robots are usually programed.

The instruction set for the 6808 CPU and the robot language commands are well documented in the Heath Corporation literature supplied with the robot and training course and will not be presented here.

DC Motor Control and Power Sources

There are 8 direct current (DC) motors used in the robot. Seven of these are stepper motors and are capable of being turned in small discrete steps. Five of the stepper motors are used in the arm, one is used for the head, and the last is used for steering the drive wheel. (Figure 1) The only non-stepper motor is a permanent magnet motor used for the drive wheel. The operation, control and programing of these motors are presented in the teaching manuals supplied with the robot. In addition, experiments are described which allow manual control of the drive motor showing how voltage supplied and polarity affect motor torque, speed and direction.

The direct current power supply for these motors as well as the control circuits is supplied by either a battery charger or the on-board gel cell batteries. The design of different batteries and their uses are also explained in the robotics course manual, as is the operation and characteristics of the battery charger used for this system.

Most industrial robots are presently operated from alternating current (AC) supplies and many use hydraulic power systems. These systems are also presented in the teaching manual, however, no experiments are described and further equipment would be desirable to reinforce the AC and hydraulic principles.

Electronic Instrumentation and Sensor Fundammentals

There are four sensors supplied with the robot; a light sensor, sound detector, motion detector and ultrasonic ranging device. Each of these sensors are described in the users manuals and the principle of operation of these sensors and, in addition, temperature and optical sensors are explained.

One of the major benefits of the HERO-1 system is that each sensor system has its own circuit board and thus is simple to isolate and explain each sensor's function separately. Teaching with these components is made even easier because they can be removed from the robot by unplugging them (Figure 1) and they will function independently in most cases. The circuits developed for these sensors are not highly sophisticated and use very few and inexpensive components. They can therefore be duplicated on prototyping boards very easily which will permit useful and inexpensive laboratory experiments. The low cost and simple nature of the HERO-1 does present a draw back in that many of the sophisticated sensors such as piezoelectric touch and camera vision systems are not available. However the HERO-1 teaching system provides an excellent foundation and introduction to these more advanced sensor systems.

Signal Conditioning-A/D and D/A

All of the sensors mentioned above will provide an analog (continuous voltage) or a digital (discrete value) signal which the CPU must recognize and then provide the appropriate response. Signal conditioning of the sensor outputs to an understandable data input for the CPU is thoroughly covered in the teaching manuals. The description of A/D and D/A conversion is highlighted with two experiments, one for sensor signal generation and the second for stepper motor control. Each of these experiments presents a detailed description of how the signals are modified and suggests many ideas for other simple sensors to the interested student.

Voice Synthesis

The HERO-1 comes equipped with a 64 phoneme sound generator with four levels of inflection for each sound. The base pitch and volume may also be manually adjusted on the speech circuit card. The speech hardware is very simple and this system is very interesting and exciting for most students. The system is easy to learn and encourages the students to program the unit to make it say their own name or any other desired sound, word or phrase. Although the voice synthesis equipment is a great deal of fun to use it has a limited practical application to industrial robotics at the present time. One serious application is to enhance the safety features of a robot by having it announce what it is going to do next or if it needs service, thus warning the surrounding personnel of upcoming actions or requirements.

The voice system is primarily to make the HERO-1 robot appear less frightening and even friendly. The 30 stored phrases in ROM allows anyone to program the unit to speak in just a few minutes and this encourages the students tremendously. The manuals fully describe the applications and the synthesizer circuit used in the HERO-1. The relative simplicity of this circuit makes it useful for further description of electronic fundamentals in a way that produces obvious and audible results.

Practical Limitations and Applications for Robots

The small size, low strength and inexact operation of the HERO-1 robot makes it nearly useless for any purpose other than education. However, it demonstrates concepts and potential applications of robotics very admirably. Specific problems encountered while programing the HERO-1 will be listed here to clarify it's limitations.

1. All seven of the robots stepper motors are susceptable to slipping if they are bumped or if they attempt to move objects that are too heavy. The arm motors should not be used to pick up objects heavier than a styroform cup, for instance.

2. The drive wheel and steering motors are not designed for long distance (> 10 feet) straight lines or to compensate for any irregularities in the floor surface.

3. The ultrasonic ranging device uses high frequency sound for ranging. This sound is produced by many things in the robot's environment, including the robot itself. This means that ranging can only be used reliably when the room is quiet and the robot is not moving.

4. Programs are not easy to edit because there is no insert feature in the monitor and no option is provided to send data to a printer. This is another potential, advanced class, experiment project. However, storage of programs on cassette tape is provided for.

5. The fully charged batteries will reliably run the robot through an active 2-3 hours, but if the "low voltage" warning is not heeded within a few minutes your programs may be lost.

These limitations apply specifically to the HERO-1 and are not typical of its industrial counterparts. The teaching manual provided by Heath Education Systems discusses robot terminology and practical applications and limitations for presently available industrial robots. Unfortunately, what is presently available today only scratches the surface of what lies ahead. The potential and applications for robots are only bounded by imagination. That is why it is so important that our engineering and technology students become exposed to the potential and possibilities for robotics to upgrade performance, efficiency and safety while lowering the costs of production.

SUMMARY

The HERO-1 Educational Training system developed by Heath Corporation is well suited for use as an educational tool for schools and industries. Although it is limited in size, strength, and accuracy to performing only insignificant tasks, it is an excellent tool for demonstrating many of the concepts required for an understanding of robotics. The unit is designed for close hands-on study and experiments. It is constructed in modules to simplify the description and understanding of each of its components. Finally, it is completely supported with users' and technical manuals, circuit diagrams, board layouts and a versatile and in depth teaching manual which covers the more general topics of robotics and industrial electronics.

References

1. Dorf, Richard C. 1983. Robotics and automated manufacturing. Restan Publishing Co. Reston, Virginia 22090.

2. Heath Company. 1982. Robotics and industrial electronics. Heath Company. Benton Harbor, Michigan 49022.

3. Heath Company. 1983. Hero robot model ET-18-technical manual. Heath Company. Benton Harbor, Michigan 49022.

4. Tanner, William R. August 1983. Robotics, their sociological implications. National Safety News. pp 51-53.

HERO 1

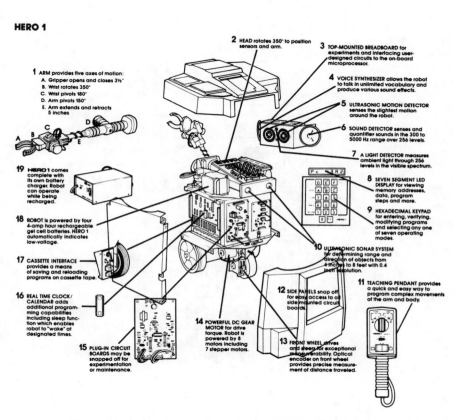

2 HEAD rotates 350° to position sensors and arm.

3 TOP-MOUNTED BREADBOARD for experiments and interfacing user-designed circuits to the on-board microprocessor.

1 ARM provides five axes of motion:
A. Gripper opens and closes 3½"
B. Wrist rotates 350°
C. Wrist pivots 180°
D. Arm pivots 150°
E. Arm extends and retracts 5 inches

4 VOICE SYNTHESIZER allows the robot to talk in unlimited vocabulary and produce various sound effects.

5 ULTRASONIC MOTION DETECTOR senses the slightest motion around the robot.

6 SOUND DETECTOR senses and quantifies sounds in the 300 to 5000 Hz range over 256 levels.

7 A LIGHT DETECTOR measures ambient light through 256 levels in the visible spectrum.

19 HERO 1 comes complete with its own battery charger. Robot can operate while being recharged.

8 SEVEN SEGMENT LED DISPLAY for viewing memory addresses, data, program steps and more.

9 HEXADECIMAL KEYPAD for entering, verifying, modifying programs and selecting any one of seven operating modes.

18 ROBOT is powered by four 4-amp hour rechargeable gel cell batteries. HERO 1 automatically indicates low-voltage.

17 CASSETTE INTERFACE provides a means of saving and reloading programs on cassette tape.

10 ULTRASONIC SONAR SYSTEM for determining range and direction of objects from 4 inches to 8 feet with 0.4 inch resolution.

11 TEACHING PENDANT provides a quick and easy way to program complex movements of the arm and body.

16 REAL TIME CLOCK/CALENDAR adds additional programming capabilities including sleep function which enables robot to "wake" at designated times.

12 SIDE PANELS snap off for easy access to all side mounted circuit boards.

14 POWERFUL DC GEAR MOTOR for drive torque. Robot is powered by 8 motors including 7 stepper motors.

13 FRONT WHEEL drives and steers for exceptional maneuverability. Optical encoder on front wheel provides precise measurement of distance traveled.

15 PLUG-IN CIRCUIT BOARDS may be snapped off for experimentation or maintenance.

Figure 1   Major Components of the HERO 1 Robot.
(Figure Furnished Courtesy of Heath Company)

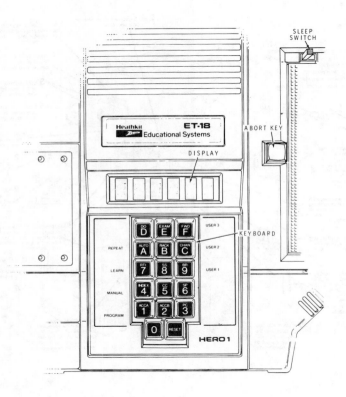

Figure 2 Hexadecimal Programing Key Pad
(Figure Furnished Courtesy of Heath Company)

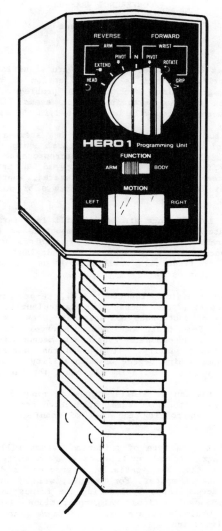

Figure 3 Motor Control and Teaching Pendant.

CONFERENCE WRAP-UP

Gerald W. Isaacs
Fellow ASAE

This conference has demonstrated that there are significant research efforts being expended worldwide aimed at the application of robotics and machine intelligence to agricultural mechanization problems. Judging from the interest expressed by the participants, it seems likely that increased research in this general area will be conducted in the future.

Interest in robotics in the classical sense has stemmed from a long history of feedback control systems used in agricultural equipment. In recent years a wide variety of electronic sensors and microprocessors have added to the versatility and effectiveness of monitoring and automatic control in farm machinery. As James Anson of Dickey-John stated, improving the efficiency of these electronic aids for the operator has a bright future in the farm equipment industry.

Higher degrees of intelligence are being built into farm machines by way of research on computer control of combine components as reported from Purdue and Clemson research. Automatic guidance of combines as reported from Japan and tractor guidance systems as reported from England and Michigan indicate the high degrees of technical feasibility if not current economic feasibility of these developments.

The application of classical robotics; i.e., re-programmable manipulators, opens new opportunities for mechanizing agricultural operations that had defied application of continuous non-selective machines like combines, mowers and grinders. Some fruit and vegetable crops like tomatoes have been bred or cultured to accommodate continuous mechanical harvest with some sacrifice of yield and quality. It appears that harvesting machines using robotics would offer a high degree of adaptability to various crops and the ability to harvest without destroying the plant for future production.

The potential application of similar types of robotic mechanisms to a wide variety of operations in agricultural production and processing should be a positive factor in controlling the cost of manufacture of future farm machinery.

It is apparent that some type of guidance systems will be needed to guide robotic arms for agricultural applications. Many current industrial applications use robots that are spatially programmed to do certain things at certain fixed points in space. For example, welding of certain spots on a car frame can be accomplished by locating the frames in the same place each time. In most proposed agricultural applications, like fruit harvesting, the objects to be manipulated are not in the same place each time or in the same spatial configuration.

The fine Australian research effort in robotic sheep shearing has met this challenge with a form of spatial reprogramming of the cutter arm based on

The author is: Gerald W. Isaacs, Chairman, Agricultural Engineering Dept., University of Florida, Gainesville, Florida.

initial measurements of the sheep, on electrical contact sensors, and programming of successive cutter passes based on the preceding one.

It is apparent that many potential robotic applications to agriculture will involve the use of vision systems to guide robotic mechanisms. Some application of vision systems is being made in manufacturing, although not yet at the technical level needed for proposed agricultural applications. Color and shape differentiation and three dimensional ranging will generally be in demand. As yet, it does not appear that anyone has fully addressed the considerable software problem of relating the output of the vision system to three dimensional action of a robotic arm.

Fortunately, much technology in vision systems has been developed for military applications by the aerospace industry and several firms specializing in industrial vision systems stand ready to assist in work on agricultural applicators.

As in the more traditional applications of automatic control to agriculture, the development of sensor technology will be a key input to advancing agricultural robotics. More basic research information on the physical properties of biological materials will be needed to relate electrical and optical properties to desired control parameters.

The potential impacts of robotics on the welfare of agricultural workers is an issue that must be addressed, hopefully with objective research and not subjective opinion. Future world market conditions will require agriculture to be competitive, thus the principal issues may be how we help labor adapt to new mechanization, not how we halt the trend toward more mechanization. As with other industries, retraining needs must be assessed.

The complex nature of potential robotics applications in agriculture indicates need for interdisciplinary research not only between agricultural engineers and biological scientists but also with other engineering disciplines as well. Teams of scientists will be needed to address the electronic, mechanical, and biological problems involved. The agricultural engineer will play a key role in clarifying the interface between the physical and biological systems.

The organizers of this conference were pleasantly surprised by the amount of research on agricultural robotics in progress around the world, although much of it is still in the planning and feasibility assessment stage. Much work remains for bench scientists and engineers before we will know the real impact of robotics on farm machinery design.

Considering the high degree of interest expressed by the conference participants and the growing interest in agricultural robotics research, the need for a future conference of this nature seems likely. The question of setting for a future conference of this nature that might be sponsored by ASAE was posed to the participants at the end of this conference. A near unanimous hand vote indicated the desirability of holding such a conference in connection with a general robotics conference and equipment exposition, as this meeting was held.

DATE DUE

	AUG 3 1 1987		
GAYLORD			PRINTED IN U.S.A.